BUILDING THE 21ST CENTURY
U.S.-China Cooperation on Science, Technology, and Innovation

Summary of a Symposium

Charles W. Wessner, Rapporteur

Committee on Comparative National Innovation Policies:
Best Practice for the 21st Century

Board on Science, Technology, and Economic Policy

Policy and Global Affairs

NATIONAL RESEARCH COUNCIL
OF THE NATIONAL ACADEMIES

THE NATIONAL ACADEMIES PRESS
Washington, D.C.
www.nap.edu

THE NATIONAL ACADEMIES PRESS 500 Fifth Street, N.W. Washington, DC 20001

NOTICE: The project that is the subject of this report was approved by the Governing Board of the National Research Council, whose members are drawn from the councils of the National Academy of Sciences, the National Academy of Engineering, and the Institute of Medicine. The members of the committee responsible for the report were chosen for their special competences and with regard for appropriate balance.

This study was supported by: Contract/Grant No. N01-OD-4-2139, TO #245, between the National Academy of Sciences and the National Institutes of Health; Contract/Grant No. DE-PI0000010, TO #15, between the National Academy of Sciences and the U.S. Department of Energy; Contract/Grant No. SB1341-03-C-0032 between the National Academy of Sciences and the U.S. Department of Commerce; Contract/Grant No. OFED-858931 between the National Academy of Sciences and Sandia National Laboratories; and Contract/Grant No. NAVY-N00014-05-G-0288, DO #2, between the National Academy of Sciences and the Office of Naval Research. Additional funding was provided by Cisco Systems, Intel Corporation, International Business Machines, the Palo Alto Research Center, the Association of University Research Parks, and Google. Any opinions, findings, conclusions, or recommendations expressed in this publication are those of the author(s) and do not necessarily reflect the views of the organizations or agencies that provided support for the project.

International Standard Book Number-13: 978-0-309-21666-1 (Book)

International Standard Book Number-10: 0-309-21666-4 (PDF)

Limited copies are available from Board on Science, Technology, and Economic Policy, National Research Council, 500 Fifth Street, N.W., W547, Washington, DC 20001; 202-334-2200.

Additional copies of this report are available from the National Academies Press, 500 Fifth Street, N.W., Lockbox 285, Washington, DC 20055; (800) 624-6242 or (202) 334-3313 (in the Washington metropolitan area); Internet, http://www.nap.edu.

THE NATIONAL ACADEMIES
Advisers to the Nation on Science, Engineering, and Medicine

The National Academy of Sciences is a private, nonprofit, self-perpetuating society of distinguished scholars engaged in scientific and engineering research, dedicated to the furtherance of science and technology and to their use for the general welfare. Upon the authority of the charter granted to it by the Congress in 1863, the Academy has a mandate that requires it to advise the federal government on scientific and technical matters. Dr. Ralph J. Cicerone is president of the National Academy of Sciences.

The National Academy of Engineering was established in 1964, under the charter of the National Academy of Sciences, as a parallel organization of outstanding engineers. It is autonomous in its administration and in the selection of its members, sharing with the National Academy of Sciences the responsibility for advising the federal government. The National Academy of Engineering also sponsors engineering programs aimed at meeting national needs, encourages education and research, and recognizes the superior achievements of engineers. Dr. Charles M. Vest is president of the National Academy of Engineering.

The Institute of Medicine was established in 1970 by the National Academy of Sciences to secure the services of eminent members of appropriate professions in the examination of policy matters pertaining to the health of the public. The Institute acts under the responsibility given to the National Academy of Sciences by its congressional charter to be an adviser to the federal government and, upon its own initiative, to identify issues of medical care, research, and education. Dr. Harvey V. Fineberg is president of the Institute of Medicine.

The National Research Council was organized by the National Academy of Sciences in 1916 to associate the broad community of science and technology with the Academy's purposes of furthering knowledge and advising the federal government. Functioning in accordance with general policies determined by the Academy, the Council has become the principal operating agency of both the National Academy of Sciences and the National Academy of Engineering in providing services to the government, the public, and the scientific and engineering communities. The Council is administered jointly by both Academies and the Institute of Medicine. Dr. Ralph J. Cicerone and Dr. Charles M. Vest are chair and vice chair, respectively, of the National Research Council.

www.national-academies.org

COMMITTEE ON COMPARATIVE NATIONAL INNOVATION POLICIES: BEST PRACTICE FOR THE 21ST CENTURY*

*As of May 2010

Project Staff *

Charles W. Wessner
Study Director

McAlister T. Clabaugh
Program Officer

David S. Dawson
Sr. Project Assistant

David E. Dierksheide
Program Officer

Peter Engardio
Consultant

Adam H. Gertz
Program Associate
(through June 2010)

Sujai J. Shivakumar
Senior Program Officer

*As of November 2010

For the National Research Council (NRC), this project was overseen by the Board on Science, Technology and Economic Policy (STEP), a standing board of the NRC established by the National Academies of Sciences and Engineering and the Institute of Medicine in 1991. The mandate of the STEP Board is to integrate understanding of scientific, technological, and economic elements in the formulation of national policies to promote the economic well-being of the United States. A distinctive characteristic of STEP's approach is its frequent interactions with public and private-sector decision makers. STEP bridges the disciplines of business management, engineering, economics, and the social sciences to bring diverse expertise to bear on pressing public policy questions. The members of the STEP Board* and the NRC staff are listed below:

*As of May 2010

STEP Staff *

Stephen A. Merrill
Executive Director

McAlister T. Clabaugh
Program Officer

David S. Dawson
Sr. Project Assistant

David E. Dierksheide
Program Officer

Charles W. Wessner
Program Director

Adam H. Gertz
Program Associate
(through June 2010)

Daniel Mullins
Program Associate

Sujai J. Shivakumar
Senior Program Officer

CONTENTS

PREFACE

Recognizing that a capacity to innovate and commercialize new high-technology products is increasingly a part of the international competition for economic leadership, governments around the world are taking active steps to strengthen their national innovation systems. These steps underscore the widely held belief that the rising costs and risks associated with new potentially high-payoff technologies, and the growing global dispersal of technical expertise, require national R&D programs to support new and existing high-technology firms within their borders.

What is the impact of these initiatives for the competitive position of the United States? In a recent report, the National Academies warned that "this nation must prepare with great urgency to preserve its strategic and economic security," adding that "the United States must compete by optimizing its knowledge-based resources, particularly in science and technology, and by sustaining the most fertile environment for new and revitalized industries and the well-paying jobs they bring."[1]

Understanding the policies that other nations are pursuing to become more innovative and to what effect is essential to understanding how the nature and terms of economic competition are shifting.[2] U.S. policymakers would benefit from knowing of the wide variety of

---

[1]National Academy of Sciences/National Academy of Engineering/Institute of Medicine, *Rising Above the Gathering Strom: Energizing and Employing America for a Brighter Future,* Washington, DC: The National Academies Press, 2007.

[2]Kent Hughes has argued in this regard that the challenges of the 21st century require new strategies that take account of new technologies, new global competitors, as well as new national priorities concerning national security and the environment. See Kent Hughes, *Building the Next American Century: The Past and Future of American Economic Competitiveness,* Washington, DC: Woodrow Wilson Center Press, 2005, Chapter 14.

innovation and competitiveness policies that leading nations have adopted. In the case of China, these innovation policies are designed to rapidly build research capacities to acquire knowledge and to transition that knowledge to national companies as a means of supporting domestic growth and employment and of building national strength.

The Overall Project

The global economy is characterized by increasing locational competition to attract the resources necessary to develop leading-edge technologies as drivers of regional and national growth. One means of facilitating such growth and improving national competitiveness is to improve the operation of the national innovation system. This involves national technology development and innovation programs designed to support research on new technologies, enhance the commercial return on national research, and facilitate the production of globally competitive products.

Here is the full Statement of Task for the project: Recognizing the importance of targeted government promotional policies relative to innovation, the Board on Science, Technology, and Economic Policy (STEP) is studying selected foreign innovation programs and comparing them with major U.S. programs. This analysis of Comparative Innovation Policy, carried out under the direction of an ad hoc Committee, includes a review of the goals, concept, structure, operation, funding levels, and evaluation of foreign programs designed to advance the innovation capacity of national economies and enhance their international competitiveness.

This analysis focuses on key areas of future growth, such as renewable energy, among others, to generate case-specific recommendations where appropriate. The Committee will assess foreign programs using a standard template, convene a series of meetings to gather data from responsible officials and program managers, and encourage a systematic dissemination of information and analysis as a means of better understanding the transition of research into products and of improving the operation of U.S. programs.

The Context of the Project

Since 1991 the STEP Board has undertaken a program of activities to improve policy makers' understanding of the interconnections among science, technology, and economic policy and their importance to the American economy and its international competitive position. The

Board's interest in comparative innovation policies derives directly from its mandate.

This mandate has previously been reflected in STEP's widely cited volume, U.S. Industry in 2000, which assesses the determinants of competitive performance in a wide range of manufacturing and service industries, including those relating to information technology.[3] The Board also undertook a major study, chaired by Gordon Moore of Intel, on how government-industry partnerships can support the growth and commercialization of productivity enhancing technologies.[4] Reflecting a growing recognition of the importance of the surge in productivity since 1995, the Board also launched a multifaceted assessment, exploring the sources of growth, measurement challenges, and the policy framework required to sustain the New Economy.[5]

The current study on Comparative Innovation Policy builds on STEP's experience to bring together leading academics, public officials, business representatives, and policy experts to better identify current trends and challenge in U.S. and foreign innovation programs.

Project Activities

To open its analysis, the study Committee held an overview symposium that drew together leading academics, policy analysts, and senior policymakers from around the globe to describe their national innovation programs and policies, outline their objectives, and highlight

---

[3]National Research Council, U.S. Industry in 2000: Studies in Competitive Performance, David C. Mowery, ed., Washington, DC: National Academy Press, 1999.

[4]This summary of a multi-volume study provides the Moore Committee's analysis of best practices among key U.S. public-private partnerships. See National Research Council, Government-Industry Partnerships for the Development of New Technologies: Summary Report, Charles W. Wessner, ed., Washington, DC: The National Academies Press, 2003. For a list of U.S. partnership programs, see Christopher Coburn and Dan Berglund, Partnerships: A Compendium of State and Federal Cooperative Programs, Columbus, OH: Battelle Press, 1995.

[5]National Research Council, Enhancing Productivity Growth in the Information Age: Measuring and Sustaining the New Economy, Dale W. Jorgenson and Charles W. Wessner, eds., Washington, DC: The National Academies Press, 2007.

their achievements.[6] Follow up symposia in Taipei and Tokyo focused on the evolution of the Taiwanese and Japanese innovation systems over the past decade. The Committee also convened a major conference in Washington that identified current trends in the Indian innovation system and highlighted the new U.S.–India innovation partnership.[7] This was soon followed by a symposium on "Synergies in Regional and National Innovation Policies in the Global Economy" held in Flanders, Belgium. This event reviewed European Union, national and regional innovation policies in Flanders, a region of Belgium, with a major university and research center with a strong commercialization record. Flanders is also home to IMEC, one of the leading microelectronics research facilities in the world and the flagship of Flemish technology policy. Also with respect to Europe, the Committee examined over a series of meetings the potential for greater U.S.-Polish cooperation in science and innovation, with particular attention to traditional energy sources (e.g., coal) and health. The Committee also held a major symposium that reviewed national strategies to foster the development of science and technology research parks around the world.[8] More recently, the Committee held a symposium in Washington, DC, on *U.S.-China Cooperation on Science, Technology, and Innovation,* which drew together speakers primarily from the U.S. and Chinese governments and academia. This was followed in June 2011 with a series of meetings in Shanghai and Beijing that included U.S. and Chinese corporate leaders and leading Chinese academic researchers. In 2010, the Committee also convened a conference on *Meeting Global Challenges: U.S.-German Innovation Policy.* A follow-up conference to this event was held in Berlin in 2011 that further compared U.S. and German approaches to support innovation and manufacturing both in terms of institutional support (e.g., by the Fraunhofer Institutes) and in specific sectors such as bio-medical, electric vehicle and solar technologies. Drawing together the information and insights from this series of meetings, the Committee will develop a

---

[6]For a summary of this conference, see National Research Council, *Innovation Policies for the 21st Century*, Charles W. Wessner, ed., Washington, DC: The National Academies Press, 2007.

[7]For a summary of this conference, see National Research Council, *India's Changing Innovation System: Achievements, Challenges, and Opportunities for Cooperation*, Charles W. Wessner and Sujai J. Shivakumar, eds., Washington, DC: The National Academies Press, 2007.

[8]The report has garnered considerable national and international attention. See National Research Council, *Understanding Research, Science, and Technology Parks: Global Best Practices-Report of a Symposium.* Charles W. Wessner, ed., Washington, DC: The National Academies Press, 2009.

consensus report that provides recommendations for U.S. innovation policy for the 21st Century.

This Workshop Summary

This report captures the presentations and discussions of the 2010 STEP symposium on U.S.-China Cooperation on Science, Technology, and Innovation. It includes an introduction highlighting key issues raised at the meeting and summary of the meeting's presentations. This workshop summary has been prepared by the workshop rapporteur as a factual summary of what occurred at the workshop. The planning committee's role was limited to planning and convening the workshop. The statements made are those of the rapporteur or individual workshop participants and do not necessarily represent the views of all workshop participants, the planning committee, or the National Academies.

Acknowledgments

On behalf of the National Academies, we express our appreciation and recognition for the insights, experiences, and perspectives made available by the participants of this meeting. We are indebted to Pete Engardio for preparing the introduction and summary of the meeting. We are also indebted to Sujai Shivakumar of the STEP staff for his work on the review of this report.

Acknowledgment of Reviewers

This report has been reviewed in draft form by individuals chosen for their diverse perspectives and technical expertise, in accordance with procedures approved by the National Academies' Report Review Committee. The purpose of this independent review is to provide candid and critical comments that will assist the institution in making its published report as sound as possible and to ensure that the report meets institutional standards for quality and objectivity. The review comments and draft manuscript remain confidential to protect the integrity of the process.

We wish to thank the following individuals for their review of this report: William Bonvillian, Massachusetts Institute of Technology, Washington, DC; Dieter Ernst, East-West Center; Patrick Keating, Stanford University; and Mu Rongping, Chinese Academy of Sciences.

Although the reviewers listed above have provided many constructive comments and suggestions, they were not asked to endorse the content of the report, nor did they see the final draft before its release. Responsibility for the final content of this report rests entirely with the rapporteur and the institution.

Alan Wm. Wolff   Charles W. Wessner

I
INTRODUCTION

INTRODUCTION

After three decades of astonishing growth, the economy of the Peoples Republic of China is nearing an important crossroad. As China's leaders themselves acknowledge, the nation's development model, based on export-led manufacturing in rich coastal provinces, cannot continue to generate sustainable, balanced growth.[1] The way forward, top leaders have stressed, is to build an economy that is driven by innovation[2]. A key question is how this transformation can take place. Can China adopt a successful innovation system directed by the state and designed to favor domestic industries? Or will China eventually adopt a more open, collaborative, and market-based system that integrates knowledge from around the world?

This crossroads for China comes as the United States faces a different kind of innovation challenge. While the United States has long been the world leader in science and new technologies, the National Academy of Sciences, in its 2007 report, *Rising Above the Gathering Storm,* warned of an abrupt loss of U.S. global leadership in science, technology, and

---

[1]On March 15, 2007, Chinese President Wen Jiabao in a press conference following the National People's Congress described China's economic model as "unstable, unbalanced, uncoordinated, and unsustainable." The remarks have been interpreted to mean that China's economy depends too heavily on fixed investment, manufacturing, and exports rather than private consumption and social equity.

[2]Former President Jiang Zemin declared innovation and high-tech industries as core to a nation's strength in the keynote address to the National Innovation Technology Conference on August 23, 1999. Current President Hu Jintao has stressed the importance of innovation in numerous speeches. Some analysts see the focus by China on innovation led growth as problematic, given how the economy is in fact advancing through stages of effective industrial supply chain collaboration and integration. See Dan Breznitz and Murphree, *Run of the Red Queen: Government, Innovation, and Globalization and Economic Growth in China,* New Haven, CT: Yale University Press, 2011.

innovation and its impact on the future prosperity of the United States.[3] This report has contributed to a growing awareness in the United States of the need to remain competitive through sustained investments in research that translate into new products, domestic industries, manufacturing, and high value employment.

COMMON CHALLENGES AND SHARED OPPORTUNITIES

China and the United States have much to gain by learning from each other as they each face their own innovation imperatives. To help advance cooperation in science, technology, and innovation, the National Academies' Board on Science, Technology, and Economic Policy (STEP) convened a symposium that brought together senior officials from China and the United States, as well as leading academics, and business people who are influential in the formation of innovation policies.[4]

This conference reflected the fact that both China and the United States share common interests in fostering science and technology to solve the challenges of economic growth, better health, and a greener environment, even as they compete in global markets. While the United States and China are the world's top two spenders on research and development, they are also by far the world's two biggest emitters of greenhouse gasses.[5] And aging populations in both countries struggle with cancer and other chronic diseases.

Indeed, a key premise of the symposium was that these and other global challenges require innovative breakthroughs, which in turn would

---

[3]The report notes that "We fear the abruptness with which a lead in science and technology can be lost—and the difficulty of recovering a lead once lost, if indeed it can be regained at all." See The National Academy of Sciences/National Academy of Engineering/Institute of Medicine, *Rising Above the Gathering Storm: Energizing and Employing America for a Brighter Economic Future,* Washington, DC: The National Academies Press, 2007.
[4]The conference, which was organized with the assistance of Cisco Systems, took place May 18, 2010, in Washington. Other sponsors of the symposium included IBM, Intel, the Palo Alto Research Center, Sandia National Laboratories, the Office of Naval Research, the Defense Advanced Research Projects Agency, the National Institute of Standards and Technology, and the National Science Foundation.
[5]China and the United States are jointly responsible for more than 40 percent of the world's greenhouse gas emissions. See *New York Times,* "China and U.S. Seek a Truce on Greenhouse Gases." Published June 7, 2009.

benefit from closer collaboration. Referring to extensive Sino-U.S. cooperation in science and technology at the commercial levels and increasingly at the university level, Deputy Assistant Secretary of State Anna Borg noted in her symposium presentation that "the United States and China are, in every sense, building a global partnership."She also identified a number of areas for closer cooperation between the two governments. Ren Weimin of China's National Development and Reform Commission agreed. Despite all the differences over which economic policies work best, Mr. Ren said, the United States and China "have a lot in common" in terms of what they must achieve. What's more, he said, the immense R&D resources and strengths of the two nations "are complementary in many respects."

SCIENCE AND TECHNOLOGY COOPERATION AND CONSTRAINTS

The National Academy of Sciences has long played a role in fostering academic and research cooperation between the United States and China. In his conference remarks Alan Wolff of the STEP Board noted that as early as 1965 the National Academy of Sciences created a committee to foster academic communication and exchanges between the two nations. After contacts were halted by the Cultural Revolution, visits resumed following the Nixon-Zhou Enlai 1972 Shanghai Communiqué.[6] In 1978, China's Ministry of Science and Technology and the U.S. National Science Foundation resumed formal cooperation. In the following year, China's Paramount Leader Deng Xiaoping and U.S. President Jimmy Carter signed the first Sino-U.S. Agreement on Science and Technology. This agreement has been extended every five years since, most recently in January 2011.[7]

Exchanges among between U.S. and Chinese scholars and technical experts have drawn this relationship closer. Some of China's brightest students attend American universities and China now is the biggest

---

[6]The Sino-U.S. Joint Communiqué, also known as the Shanghai Communiqué, was issued on February 28, 1972, following President Richard Nixon's historic seven-day trip to China.
[7]There have also been challenges to extending S&T cooperation with China in space technologies. See *Politico*, February 12, 2011, "House continuing resolution would bar NASA from China ties."

source of foreign students in U.S. science and engineering programs.[8] A surge of investment in research facilities in the 1990s by American corporations in the mainland further bound the two science and engineering communities.

Over the past decade, the United States and China have signed some 50 cooperative agreements in fields such as agriculture, energy resources, the environment, and basic science, involving nearly every Chinese government agency.[9] Speaking at the conference, Yang Xianwu of China's Ministry of Science and Technology said that "Cooperation with the U.S. has always been our priority."

At the same time, the relationship between China and the United States has also been characterized by frictions and competing agendas. U.S. companies frequently complain about China's weak protection of intellectual property rights.[10] And officials and business leaders from the United States have joined those in Europe and India in objecting to what they see as discrimination against foreign companies stemming from Chinese industrial policies and a growing focus on *zizhu chuangxin,* widely translated as "indigenous innovation."[11] Because domestic

---

[8]Institute for International Education, "International Student Enrollments Rose Modestly in 2009/10, Led by Strong Increase in Students from China." 2011 Press Release. Access at <*http://www.iie.org/en*>.

[9]Perhaps the most far-reaching partnership is in energy, where the two governments signed the Protocol on Sino-U.S. Joint Research Center for Clean Energy. Each country will invest $15 million in the new program and assign its own staff. The center will facilitate joint research and development in an array of clean energy technologies. The agreement to establish the center was announced July 15, 2009 by U.S. Energy Secretary Steven Chu, Chinese Minister of Science Wan Gang, and Administrator of National Energy Administration Zhang Guo Bao.

[10]For an example of U.S. industry complaints, see John Neuffer, "China: Intellectual Property Infringement, Indigenous Innovation Policies, and Frameworks for Measuring the Effects on the U.S. Economy," written testimony to the United States International Trade Commission Investigation No. 332-514 Hearing on behalf of the Information Technology Industry Council, June 15, 2010.
(<*http://www.itic.org/clientuploads/ITI%20Testimony%20to%20USITC%20He aring%20on%20China%20%28June%2015,%202010%29.pdf* >).

[11]The Chinese policy for indigenous innovation, *zizhu chuangxin,* was introduced in a 2006 state-issued report, "Guidelines on National Medium- and Long-Term Program for Science and Technology Development." Some Chinese sources describe the policy as encouraging research institutes and universities to conduct innovative research and create new intellectual property to meet

companies are favored in government purchases[12]—which account for the lion's share of spending on infrastructure and information technology—foreign companies say selling their products in China is increasingly difficult.[13]

Meanwhile, some Chinese officials who spoke at the symposium cited overly restrictive U.S. export controls rules on certain "dual use" technologies as needlessly blocking U.S. sales of some high-performance computers, advanced semiconductor manufacturing equipment, and numerically controlled machine tools to China.[14] They noted that these curbs have, for example, prevented Chinese companies from buying

---

national demands and to build up China's innovation capacity. Many foreign firms operating in China however believe that the policy would seek to transfer their patents and other intellectual property to China in order to compete for technology and equipment procurement by the Chinese government. For an analysis of China "indigenous innovation" policy, see Adam Segal, "China's Innovation Wall: Beijing's Push for Homegrown Technology," *Foreign Affairs*, September 28, 2010. Describing the impact of this policy, Segal notes that "In 2009, for example, China's government, a massive consumer of high-tech products, announced that in order to be a recognized vendor in the government's procurement catalog, a company would have to demonstrate that its products included indigenous innovation and were free of foreign intellectual property."

[12]China's 15-year plan for science and technology says the government should practice a "first-buy policy for major domestically made high-tech equipment and products that possess proprietary intellectual property rights." See Sec VIII, 3 of "The National Medium- and Long-Term Program for Science and Technology Development (2006-2020): An Outline," pg. 54, State Council of China.

[13]In a March 2010 survey by the American Chamber of Commerce in Beijing, 37 percent of U.S. information technology companies said they would lose sales because of "indigenous innovation" policies, leading the Obama Administration to take up this practice at the highest levels. In his May 25, 2010, press briefing in Beijing, Timothy Geithner said that Chinese leaders had expressed "principles of nondiscrimination" regarding China's indigenous innovation policy and that this represented significant progress.
(<*http://www.state.gov/secretary/rm/2010/05/142198.htm*>). In June 2011, China's Ministry of Finance announced that it is scrapping certain rules designed to foster "indigenous innovation. See *Wall Street Journal*, July 1, 2011, "China Plans to Ease Rules That Irked Companies."

[14]The Bureau of Industry and Security of the Commerce Department enforces Export Administration Regulations (EAR), which restricts exports of items that have both commercial, military, or proliferation applications. American companies often complain they lose billions of dollars in business in China to other nations that do not have these restrictions. See AmCham-China, op. cit.

equipment from Applied Materials—a U.S. firm—to mass manufacture the current-generation 300 mm silicon wafers. The Chinese participants at the symposium also cited U.S. immigration rules as another irritant. After the 2001 terrorist attacks, they said, it has become harder for Chinese citizens to obtain U.S. entry visas is a timely manner.

BACKGROUND ON CHINA'S INNOVATION SYSTEM[15]

High-Level Commitment and Growth

China's modern innovation system is rooted in the reforms of the late 1970s. At that time, China's scientific community and university system had been decimated by the Cultural Revolution. Communist Party leaders such as Marshall Nie Rongzhen and Deng Xiaoping argued that science and technology were vital to modernize China's military and meet basic social needs.[16] In 1975, Premier Zhou Enlai named science as one of the Four Modernizations. After he assumed power, Deng advocated at a National Science Conference in 1978 that scientific institutes be run by administrators and scientists, not party cadres. "Without the rapid development of science and technology it will become impossible to build the national economy," Deng declared.[17]

More recently, in a report to the 17th National Congress of the Communist Party of China, President Hu Jintao stated that "Innovation is the core of our national development strategy and a crucial link in enhancing the overall national strength." This high level commitment has been backed by a sharp rise in China's commitment to R&D spending—from a six percent share of global R&D spending in 1999 to an estimated 12.2 percent share in 2010.[18]

Over the past 15 years, China has launched many initiatives to boost science, develop high-tech industries, and reduce its dependence on

---

[15]The text in this section provides a brief historical background on the evolution and current challenges facing China's innovation system. While not directly based on the discussions held at the symposium, the text here provides additional context to the conference discussions.

[16]See Evan A. Feigenbaum, *China's Techno-Warriors: National Security and Strategic Competition from the Nuclear to the Information Age*, Stanford: Stanford University Press, 2003.

[17]Deng Xiaoping address at the First National Science Congress, 1978.

[18]Battelle, *R&D Magazine*. December 2009.

foreign technologies.[19] The 973 Program, for example, supports 175 chief scientists focusing on "strategic needs," such as agriculture, energy, information, and health.[20] The 863 Program, better known as the State-High Tech Development Plan, is aimed at easing China's dependence on imported advanced technologies and is credited with leading to the development of China's Shenzhou spacecraft and Loongson computer processor. The Torch Program, meanwhile, promotes development of high-technology industrial zones.[21]

"Indigenous innovation" has become a top priority in the past five years. As "guiding principles for science and technology undertakings," China's National Medium- and Long-Term Program for Science and Technology Development for 2006 to 2020 lists "indigenous innovation, leapfrogging in priority fields, enabling development, and leading the future."[22] The document says that "in areas critical to the national economy and security, core technologies cannot be purchased," and that China must "master core technologies in some critical areas, own proprietary intellectual property rights, and build a number of internationally competitive enterprises." The plan calls for boosting China's gross R&D spending to 2.5 percent of GDP by 2020, for science

---

[19]For a comprehensive review of China's innovation policies, see Micah Springut et al., "China's Program for Science and Technology Modernization: Implications for American Competitiveness." Prepared for the U.S.-China Economic and Security Review Commission. January 2011. There is also a growing literature by Chinese scholars on policy issues related to innovation. See, for example Zheng Liang and Lan Xue, "The evolution of China's IPR system and its impact on the patenting behaviours and strategies of multinationals in China," *International Journal of Technology Management.* 51(2/3/4), 2010. See also, Shulin Gu and Bengt-Åke Lundvall, "Policy learning as a key process in the transformation of the Chinese Innovation Systems," in Bengt-Åke Lundvall, Patarapong Intarakumnerd, and Jan Vang, Eds., *Asian innovation systems in transition,* Edward Elgar Publishing Ltd, 2006.

[20]The National Basic Research Program, also known as the 973 Program, was approved by the central government in June 1997 and administered by the Ministry of Science and Technology. For an explanation in English of the program, see <*http://www.973.gov.cn/English/Index.aspx*>.

[21]For a concise explanation of Chinese innovation policies over the past decade, see Can Huang, Celeste Amorim, Mark Spinoglio, Borges Gouveia and Augusto Medina, "Organization, Programme and Structure: An Analysis of the Chinese Innovation Policy Framework," *R&D Management* 34(4), 2004. (<*http://xcsc.xoc.uam.mx/apymes/webftp/documentos/biblioteca/analysis%20of %20the%20Chinese%20innovation%20policy.pdf*>.)

[22]For a U.S. perspective of the impact of "indigenous innovation," See Adam Segal, op. cit. See also Breznitz and Murphree, op. cit.

and technology to account for 60 percent of the economy, and cutting dependence on imported technology to 30 percent.[23]

China's Innovation Challenges

This strong emphasis by its leaders has led to some remarkable progress: Investment in research and development, patent filings, output of published scientific papers, exports of high-technology electronic products, and engineering and science graduates with advanced degrees all have risen rapidly over the past decade.[24]

However, China's output of new technologies and breakthrough product remains weak. An extensive study of China's innovation system by the Organization for Economic Co-Operation and Development and the Chinese Ministry of Science and Technology concluded that the heavy investments have "yet to translate into a proportionate increase in innovation performance." The report faulted "deficiencies in the current policy instruments and governance for promoting innovation."[25]

A World Bank study of Chinese enterprises reached similar conclusions.[26] A survey of nearly 300,000 of industrial enterprises found 53 percent of large enterprises, 86 percent of medium-sized, and 96 percent of small in 2004 through 2006 did not have continuous research and development. As a result, they don't own core technologies and rely

---

[23]State Council of China, "National Medium- and Long-Term Program for Science and Technology Development, 2006-2020," (<*http://webcache.googleusercontent.com/search?q=cache:y800l0iQlS8J:www. cstec.org/uploads/files/National%2520Outline%2520for%2520Medium%2520a nd%2520Long%2520Term%2520S%26T%2520Development.doc+china+Natio nal+Medium-+and+Long- Term+Program+for+Science+and+Technology&cd=18&hl=en&ct=clnk&gl= us&client=firefox-a*>.)

[24]See National Research Council, *The Dragon and the Elephant, Understanding the Development of Innovation Capacity in China and India.* S. Merrill ed., Washington, DC: The National Academies Press, 2010.

[25]*OECD Reviews of Innovation Policy: China,* Organization for Economic Co-Operation and Development, September 2008. OECD Publishing. (<*http://www.oecd.org/dataoecd/7/45/41270116.pdf*>.)

[26]Chunlin Zhang, Douglas Zhihua Zeng, William Peter Mako, and James Seward, *Promoting Enterprise-Led Innovation in China*, Washington, D. C.: The International Bank for Reconstruction and Development/The World Bank, 2009. (<*http://siteresources.worldbank.org/CHINAEXTN/Resources/318949- 1242182077395/peic_full_report.pdf*>.)

"upon factors other than innovativeness" to compete globally. The study described the system as one of "manufacturing without innovation."[27]

In addition to low corporate R&D investment, these studies cite a number of other reasons for China underachievement in innovation. They include weak intellectual property protection, shortages of capable and skilled personnel, an over-emphasis on export manufacturing of commodity goods, and weak linkages between government-funded research institutions and the private sector.[28]

China's Innovation Agenda

Speaking at the conference, some Chinese officials said that they were aware of these shortcomings. In his presentation, Yang Xianyu of the Ministry of Science and Technology said that Chinese businesses have been slow to invest in R&D and that "a key challenge is to transform China's economic development pattern so that it is driven by innovation."

To spur investment in innovation, the government is offering generous tax incentives to companies in "high-priority" sectors and that meet certain R&D investment benchmarks, Mr. Yang said. For every *renmenbi* spent on R&D, they get 1.5 *renmenbi* in tax credits.[29] The government also is establishing more small-business incubators in science and technology parks and training centers for entrepreneurs. It is boosting funding for national laboratories, engineering centers, and university science parks.

---

[27]The economies of previously emerging economies have followed a comparable pattern of manufacturing without innovation. For analyses of these cases, see, for example, Linsu Kim, *Imitation to Innovation; The Dynamics of Korea's Technological Learning*, Boston: Harvard Business School Press, 1997, at for example, pp. 192-213, 234-243. See also Glen Fong, who has shown a series of stages through which innovative economies must move – see Glenn R. Fong, "Follower at the Frontier: International Competition and Japanese Industrial Policy," *International Studies Quarterly*, 42(2), 1998.

[28]See Denis Fred Simon and Cong Cao, *China's Emerging Technological Edge: Addressing the Role of High-End Talent,* Cambridge: Cambridge University Press, 2009.

[29]Tax incentives for R&D in China is vary with the location of the investment and the type of technology in use. In addition, provincial and local governments often provide additional tax advantages for corporate R&D. The PRC Government's R&D tax credit is permanent and offers businesses a tax deduction of 150 percent, if R&D spending increases 10 percent over the previous year. See <*http://www.investinamericasfuture.org/*>.

Easing China's dependence on imported technologies and strengthening "indigenous innovation" are high policy priorities, Mr. Yang said. China remains committed to international collaboration as a vehicle to "absorb innovation" that can be adapted to "Chinese conditions," he said. Beijing is focusing resources on 16 science and technology areas identified in the 2006-2020 Plan, such as nano-materials and semiconductors, in which the nation should become more self-sufficient.

IN THE UNITED STATES, A RENEWED FOCUS ON INNOVATION

America's innovation system also is amid a reassessment. The United States remains the world leader in patents, R&D investment, scientific papers and other standard benchmarks of innovation. The United States also still produces many high-technology start-ups, many of which like Google and Microsoft have rapidly grown to become world leaders. As the National Academies' 2007 report, *Rising Above the Gathering Storm*[30] explained, however, there is mounting concern that America's global competitiveness is eroding, largely due to underinvestment in scientific research, falling math and science skills, and engineering talent shortages.

Focus on Manufacturing and Jobs

The deep recession triggered by the 2008 financial crises brought another concern into focus: That U.S. inventions are not creating enough high-paying U.S. jobs and new globally competitive industries.[31] As X/Seed Capital founding partner Michael Borrus noted in his symposium presentation, the United States still is, for example, the leading source of innovation in solar cells and modules, but most of the manufacturing is ending up in China.

The U.S. federal government and state governments have sought to stimulate development of new domestic industries through loans and grants to manufactures of electric cars, lithium-ion batteries, and thin-

---

[30]National Academy of Sciences/National Academy of Engineering/Institute of Medicine, *Rising Above the Gathering Storm: Energizing and Employing America for a Better Economic Future*, op. cit.
[31]See Pete Engardio, "Can the Future be Built in America? Inside the U.S. Manufacturing Crisis," *BusinessWeek* September 21, 2009.

film solar cells.[32] As Ginger Lew of the White House National Economic Council noted in her symposium remarks, the Obama Administration is taking steps to coordinating support for regional economic clusters across a number of federal agencies.[33]

Importance of U.S.-China Cooperation

Recognizing the highly globalized nature of research and innovation in the 21st century, the United States is also seeking to collaborate more closely with China on areas of common challenges and shared interest. As highlighted below, participants at the conference described a number of shared challenges and potential areas for cooperation and mutual learning, including in the development and commercialization of renewable energy and information and communications technologies, the development of research parks and innovation clusters, university reform, and addressing the shared challenges of medical research. Lastly, participants also described some challenges to closer U.S.-China cooperation in high-technology research and commercialization.

COOPERATION ON RENEWABLE ENERGY INNOVATION

China's Renewable Energy Imperative

Speaking at the symposium, Ren Weimin of the National Development and Reform Commission and Kristina Johnson, then Under Secretary at the Department of Energy, described how both nations can gain through collaboration in renewable energy innovation. Renewable energy innovation is one area where both China and the United States stand to gain from collaboration. As explained in his presentation, Ren Weimin said that China faces enormous challenges meeting the future energy needs of its rapidly developing economy. Over the past five years, China's energy consumption has nearly doubled, to 3.1 million

---

[32]See National Research Council, *Building the U.S. Battery Industry for Electric-Drive Vehicles: Progress, Challenges, and Opportunities*, Charles W. Wessner, Rapporteur, Washington, DC: The National Academies Press, forthcoming. See also National Research Council, *The Future of Photovoltaic Manufacturing in the United States*, Charles W. Wessner, Rapporteur, Washington, DC: The National Academies Press, 2011.
[33]See symposium presentations by Ginger Lew of the National Economic Council and U.S. Energy Under Secretary Kristina Johnson, in the Summary of Presentations chapter of this volume.

tons of coal equivalent, he noted. Over the next four decades, energy use is projected to more than double again. Currently, China relies almost entirely on fossil fuels, especially domestically mined coal, to generate electricity. "Against this background, renewable energy is our inevitable choice," he said.

China has ambitious targets for clean energy. Beijing wants non-fossil fuels to account for 15 percent of consumption by 2020, 20 percent by 2030, and one-third by 2050.[34] That compares to 8.3 percent now. Like the United States, China hopes to fill these energy requirements with a mix of solar, wind, hydro, nuclear, bio-fuels, thermal, and clean coal.

China has enormous untapped resources in most of these renewable sources, Mr. Ren said. China also is the world's biggest manufacturer of photo-voltaic cells and panels and has the fastest-growing installed base of wind generators. The country is producing enough biogas to provide fuel to 80 million rural people and enough geothermal to provide 600,000 people with heated water. This gives China "a solid foundation for developing renewable energy," he said.

The problem is that, "from the perspective of price, renewable energy is very expensive," Mr. Ren said. Power generated by coal mined in Xinjiang Province costs the equivalent of 3.4 cents per kilowatt. Wind power costs 7 to 9 cents and solar power at least 19 cents.

China remains far from making wide deployment of renewable energies commercially viable, Mr. Ren said. Shortcomings include an inadequate "industrial system," policy coordination, "market monitoring mechanisms," and legal frameworks, he said. China's weakness in technical innovation and basic research are other handicaps. While China's solar- and wind-power equipment is large and growing fast, he noted that manufacturers must import key technology, equipment, and raw materials. Therefore, Mr. Ren said China is developing a "comprehensive policy and institutional framework" for renewable energy. "Economic and industrial policy should be compatible with energy policy," he said.

---

[34]State Council of China, "National Medium- and Long-Term Program for Science and Technology Development, 2006-2020," (<*http://webcache.googleusercontent.com/search?q=cache:y800l0iQlS8J:www. cstec.org/uploads/files/National%2520Outline%2520for%2520Medium%2520a nd%2520Long%2520Term%2520S%26T%2520Development.doc+china+Natio nal+Medium-+and+Long-Term+Program+for+Science+and+Technology&cd=18&hl=en&ct=clnk&gl= us&client=firefox-a*>).

America's Renewable Energy Push

In her presentation, then Energy Under Secretary Kristina Johnson noted that the United States also has recently launched a number of initiatives to cut greenhouse gas emissions by 83 percent by 2050. Other goals include doubling renewable electricity generation and advanced energy manufacturing by 2012. While the United States wants to demonstrate global leadership in energy science and technology, its approach to innovation will be "open and collaborative," "By working together, we can leverage our comparative advantages in innovation and address this global climate challenge," she said.

Some 70 percent of U.S. electricity comes from fossil fuels. The United States plans a "fundamental shift" in the way it generates power, Dr. Johnson said. It will expand commercial nuclear-power, install carbon-capture and storage technologies in coal-fired plants, and increase renewable energy. Other priorities are to de-carbonize transportation, which consumes 29 percent of U.S. energy, and improving energy efficiency in buildings, which consume 40 percent.

The Obama Administration has sharply boosted spending on renewable-energy technologies. It devoted $80 billion under the American Recovery and Reinvestment Act[35] to clean-energy projects, with half of that going to the Department of Energy. Private investors mobilized another $150 billion, Dr. Johnson said.

The DoE's heavy emphasis on basic research also has shifted, Dr. Johnson said. Three-quarters of Recovery Act funds are for projects aimed at deploying new technologies. The DoE invested $3.4 billion to develop next-generation vehicles and fueling infrastructure, for example. This is on top of the $8.4 billion extended in the Advanced Technology Vehicle Manufacturing Loan Program.[36] Companies are using the funds to build three new electric-vehicle plants and 30 battery and electric-vehicle component plants. Another $600 million is going to 19 pilot, demonstration, and commercial-scale bio-refineries for new fuels.

---

[35]The American Recovery and Reinvestment Act of 2009, HR 1, was signed by President Barack Obama on Feb. 17, 2009. It funded some $780 billion in programs to stimulate the U.S. economy.
[36]The Advanced Technology Vehicle Loan program is administered by the Department of Energy. First funding of grants, loans, and other incentives to makers of automobiles and auto parts to support development and manufacturing of advanced vehicles was provided under Section 136 of the Energy Independence and Security Act of 2007.

The bio-fuels initiative illustrates the DoE's comprehensive new for innovation strategy. It is setting up Energy Frontier Research Centers to focus on scientific discovery. The Advanced Research Project Agency for Energy (ARPA-E) funds applied research projects to develop new fuels. Another initiative, to establish energy-research hubs,[37] aims to accelerate large-scale commercial deployment of new bio-fuels. The DoE also is creating a regional innovation hub for energy-efficient building technologies.

The federal government is forging partnerships with companies, university, regional governments, and foreign research institutes to achieve these goals. China is an important partner. The U.S.-China Clean Energy Research and Development Center, announced in July, shows how close the relationship is becoming. Each nation will invest $75 million over five years for joint research on energy-efficient buildings, vehicles, and carbon capture and sequestration for coal.[38]

Collaborating on Renewable Energy Research and Commercialization

The National Renewable Energy Laboratories (NREL) in Golden, Colorado, also is engaged in a broad and deepening relationship with China. Somewhat unique among DoE labs, NREL's Robin L. Newmark explained in her presentation, it studies the economic and policy issues related to renewable energies as well as the technologies themselves.

A flurry of new projects was sparked by an umbrella agreement negotiated through the U.S.-China Strategic Economic Dialogue, Ms. Newmark said. NREL's collaborations with Chinese companies, research institutes, and government agencies now range from long-range planning of wind power to commercializing specific bio-fuels.

At the macro level, NREL is involved with two new Sino-U.S. research centers. One analyzes China's national potential in wind power and the technical, economic, and logistical issues of connecting large

---

[37]For explanations of recent Department of Energy innovation initiatives, see Kristina Johnson presentation in upcoming book National Research Council, *Clustering for 21st Century Prosperity*, Charles W. Wessner, Rapporteur, Washington, DC: The National Academies Press, forthcoming.

[38]See Department of Energy, *U.S.-China Clean Energy Cooperation: A Progress Report by the U.S. Department of Energy*, January 2011. The DoE report that highlights the areas and status of U.S.-China clean energy cooperation.

(<*http://www.pi.energy.gov/documents/USChinaCleanEnergy.PDF*>.)

wind farms to China's power grid. The other center explores similar challenges with solar power.

The partnership in bio-fuels spans the innovation chain, involving several DoE and Department of Agriculture labs, Chinese research institutes, and mainland companies such as SinoPec, PetroChina, CNOOC, and COFCO. NREL is helping study economic and technical solutions for supplying bio-fuel feed stocks other than food sources, for example. Another project, with Beijing's Tsinghua University and PetroChina, focuses on the process of breaking down bio-materials so they can be converted into fuel, while a partnership with the Chinese Academy of Sciences seeks to develop biodiesel from algae and plant oils.

The lab is helping Chinese companies commercialize renewable energy technologies as well. NREL is partnering with ENN Group Co. to develop large solar cells to be manufactured with technology from Applied Materials, for example. Ms. Newmark said she sees "enormous opportunities" for innovation that will benefit both countries and that she expects "rapid growth" in such bilateral partnerships.

Comparing Policies on Innovation Parks and Clusters

Research parks are increasingly seen as an effective tool to create dynamic clusters of research, manufacturing, and services that encourage innovation and foster economic growth.[39] U.S. and Chinese speakers at the conference offered contrasting approaches to the development of these innovation clusters.

While policies and investments at the national level are important, innovation takes place at a local scale, said Ginger Lew of the White House National Economic Council. She noted that most innovation zones in the United States are initiated by consortia of city and state governments and business, community, and educational leaders as an economic development tool. While many innovation clusters have

---

[39]Research Parks, often also known as Science and Technology Parks, are made up of a collection of buildings and facilities that are dedicated to research and development. See National Research Council, *Understanding Research, Science, and Technology Parks: Global Best Practices: Report of a Symposium.* Charles W. Wessner, ed., Washington, DC: The National Academies Press, 2009. Research parks that foster dense networks of trust and cooperation among the small and large businesses, research institutes, and other park participants can develop into clusters of innovative activity. See National Research Council, *Growing Innovation Clusters for American Prosperity*, Charles W. Wessner, Rapporteur, Washington, DC: The National Academies Press, forthcoming.

benefitted from substantial federal investments in nearby research universities and national laboratories, there has until recently been no coordinated federal support to encourage their development. Whereas many U.S. agencies had been operating in "silos," Ms. Lew noted that a key feature of the Obama Administration's strategy is to coordinate programs of various federal agencies to support "holistic, integrated solutions to building regional economies."

Initiatives launched in the United States over the past year include the Energy Regional Innovation Clusters (ERIC) program, in which the DoE is leading six other federal agencies to help U.S. regions develop innovation zones.[40] The U.S. Department of Agriculture is awarding grants to 12 rural communities to create new industries based on their traditional ones. The Small Business Administration will support five regional initiatives to commercialize new technology. And in May 2010, the Department of Commerce said it will award $12 million in grants to six U.S. teams with "the most innovative ideas to drive technology commercialization and entrepreneurship in their regions," Ms Lew explained.

In all, President Obama's budget for Fiscal Year 2011 authorized more than $300 million in new funding for federal agencies to assist regional innovation cluster initiatives, Ms. Lew said. The current version of the America COMPETES Act also includes provisions to promote clusters.[41]

In his conference remarks, Carl Dahlman said that China has made "dramatic progress" in setting up all kinds of science and innovation parks. Introducing Lou Jing of China's Ministry of Education, he added that "I think we are very lucky to have with us one of the key people behind that." In her symposium remarks, Ms. Lou noted that China has developed numerous innovation clusters. As examples, she cited the Zhongguancun and Shandi districts in Beijing, the high-tech development zone in Shanghai, and the science and technology parks and research centers at universities and in provinces around the country.

---

[40]The first Energy Regional Innovation Cluster is to focus on clean-energy technologies used in buildings. For details, see the Funding Opportunity Announcement for Fiscal Year 2010 on the DoE Web site. See <*http://www.energy.gov/hubs/documents/ERIC_FOA.pdf*>.

[41]The America COMPETES Reauthorization Act of 2010 (H. R. 5116) passed the House of Representatives on May 28, 2010. It revises the original America COMPETES Act (P.L. 110-69). Despite being enacted on Aug. 9, 2007, funding was never appropriated.

Indeed, China has sought to develop large-scale research parks to accelerate the development of China's industrial base in areas such as electronics and information technology, new materials, and bio-medicine.[42] Although size is not necessarily a measure of success, the scale of China's 54 state-level science and technology industrial parks is remarkable. (See Box A.)

Yang Xianwu of the Ministry of Science and Technology noted in his symposium remarks that the Chinese government is supporting the development of innovation clusters by financing the establishment of laboratories, engineering centers and large science facilities. It is aiding projects that can serve as catalysts, he explained, such as university science parks, high-tech industrial parks, and innovation centers. "We're learning from the experience of Finland and America's Silicon Valley by establishing a large number of incubation centers to help scientists transform their research results and open their own small and medium-sized enterprises," he said.

COOPERATION ON 21ST CENTURY UNIVERSITIES

Participants in the conference, as we see below, observed that both the United States and China are relying more on universities, which traditionally have focused on education and research, to serve as anchors for regional innovation clusters and catalysts of economic development.

To get across the important role U.S. universities play in the economy, University of Maryland at College Park President C. D. Mote offered some statistics from his state. Every dollar Maryland spends on his university, he said, generates $8 in economic activity. The university also raises $35 in development resources for small Maryland businesses. Over 25 years, he said, each dollar in state investment has generated $200 worth of goods and services.

---

[42]Kazuyuki Motohashi and Xiao Yun, "China's innovation system reform and growing industry and science linkages." *Research Policy* 36: 1251-1260, 2007.

BOX A
Research Parks in Comparative Perspective—an Issue of Scale

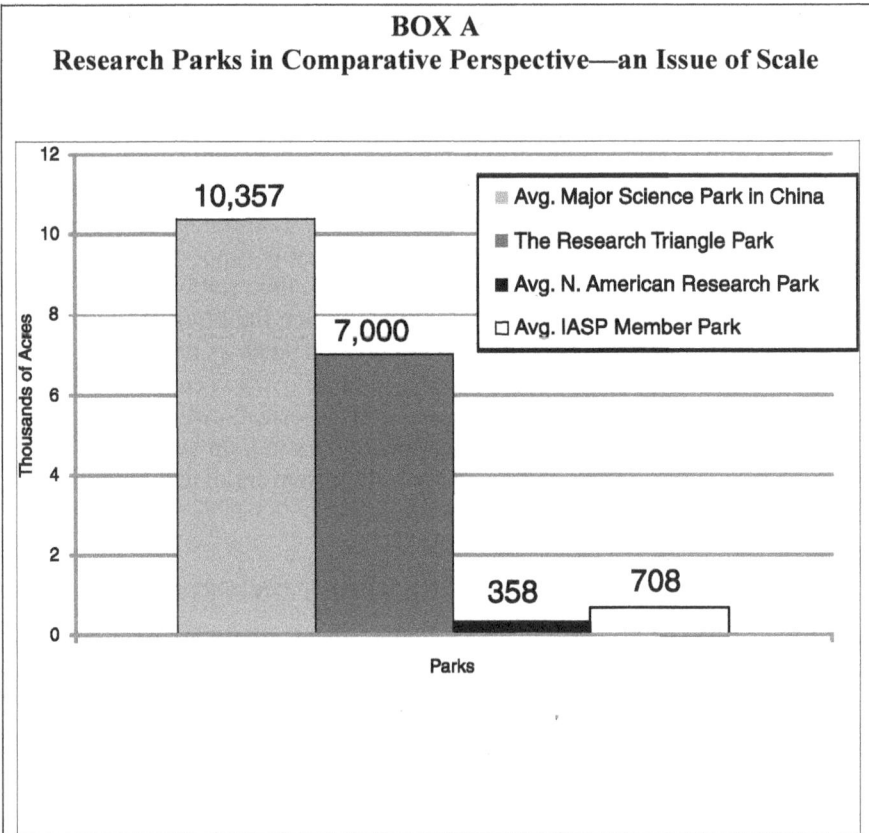

FIGURE A-1 Relative sizes of U.S. and Chinese research parks.
SOURCE: Presentation by Richard Weddle in *National Research Council,*
Understanding Research, Science and Technology Parks: Global Best Practices,
C. Wessner, ed., Washington DC: National Academies Press, 2009.
NOTE: "Average North American Research Park" data are from
"Characteristics and Trends in North American Research Parks: 21st Century
Directions," commissioned by AURP and prepared by Battelle, October 2007;
"Average IASP Member Park" data are from the International Association of
Science Parks annual survey, published in the 2005-2006 International
Association of Science Parks directory.

China has made rapid progress in higher education, Carl Dahlman of
Georgetown University observed. Enrollment rates have risen from 2
percent in 1980 to 23 percent today. Now, China has more people in
universities, 25 million students, than the United States, with 17 million.
China spends more of its R&D money in universities than most other
nations, including the United States, he said. China also has made

"dramatic progress" in setting up science and innovation parks, where universities are transferring knowledge to the private sector, Dr. Dahlman said.

China's Commitment to University Growth

Universities are central to China's strategy to build a "system of innovation with Chinese characteristics," Lou Jing of the Ministry of Education's Department of Science and Technology said in her presentation.[43]

China's R&D infrastructure is heavily concentrated on campuses. Sixty percent of China's "national pilot laboratories" and nearly two-thirds of its 140 "national key laboratories" are based at universities, Ms. Lou noted. So are 26 national engineering laboratories and 110 National Engineering Research Centers. There are 76 science parks with connections to more than 110 universities, she said. Universities are in charge of some 80 percent of research under National Science Foundation general programs, and 40 percent of national high-technology research-and-development programs.

Chinese universities are assuming bigger roles in innovation. Funding for applied research is growing 20 percent annually, she noted. Universities produce more than one-third of Chinese patents for inventions and 60 percent of published science and engineering papers.

Other key elements of this ecosystem are the Chinese Academy of Sciences, the Chinese Academy of Social Sciences, research institutes specializing in economics and social development, and the Chinese research organizations of multinationals such as IBM and Cisco. Ms. Lou said the government's vision is for a technological innovation system that is "business-based, market-oriented and that integrates industry, academia, and research."

The first mission of universities is "to serve as an engine or driver of a country's core competitiveness," Ms. Lou said. To do so, there must be closer collaboration between academia, industry, and research institutes, she said. The government also wants to "markedly raise competitiveness and the quality of higher education," she said.

---

[43]For a discussion of productivity growth at Chinese Universities, see Ying Chu Ng and Sung-ko Li, "Efficiency and productivity growth in Chinese universities during the post-reform period." *China Economic Review* 20, 2009.

Universities in America's Innovation System

Universities have played a significant role in the U.S. innovation system since the Civil War, when the federal government began allotting land to each state to establish institutions to teach agriculture and engineering, National Academy of Engineering President Charles Vest explained in his presentation. That role expanded after World War II, when the federal government set up a system to fund basic research at universities, and in 1980, when the Bayh Dole Act allowed universities to commercialize intellectual property generated by federally funded research. "This started a very different and increased relationship of universities to the private sector," explained Dr. Vest, a former president of the Massachusetts Institute of Technology.

BOX B
A Complex Innovation System

In his conference presentation, Charles Vest of the National Academy of Engineering observed that the development of innovative products is increasingly the result of knowledge that flows back and forth among complex, inter-linked, and often ad-hoc "innovation ecosystems" at universities, corporations, government bodies, and national laboratories. The so-called U.S. innovation system "frankly is not really a system," he said. "It is not designed or planned very explicitly."

The innovation process involving government, universities, and industry has historically been "very decentralized, very loosely organized, and highly entrepreneurial," Dr. Vest said.[44] It also tends to vary from region to region. But it has worked remarkably well at producing commercial products, processes, and services. An estimated 60 percent of America's economic growth has been attributed to technological innovation, and the system has produced such "earth-shaking" advances as computing, the laser, the World Wide Web, financial engineering, and much of modern medicine.

---

[44]The U.S. innovation system is characterized by both decentralization as well as strong networks of collaboration. Vernon Ruttan has noted that nearly all the major world innovation waves of the second half of the 20th century were characterized by government initiated linkages across the innovation system. See, Vernon W. Ruttan, *Is War Necessary for Economic Growth, Military Procurement and Technology Development.* Oxford University Press, 2006.

None of these breakthroughs "were explicitly planned or envisioned in advance," Dr. Vest observed. Nor were some of America's most important innovation clusters, such as Silicon Valley or Boston's Route 128. The question now is how to adapt the U.S. innovation system at a time when the venture-capital industry has become more averse to risk and to deal with enormous challenges such as energy, climate change, food, and water, Dr. Vest said.

How the University of Maryland Drives Growth

The University of Maryland at College Park illustrates the broad range of ways in which a university can impact the innovation economy—locally, regionally, nationally, and even internationally, university President Mote said in his presentation.

Dr. Mote explained that "the spirit of entrepreneurship is embedded into the infrastructure of the university."[45] The University of Maryland has a special dormitory for student entrepreneurs, for example, that spawns an average of 17 start-up companies a year. Another program works with community colleges to nurture entrepreneurs in their 30s and 40s. The university's engineering school has run the Maryland Technologies Enterprise Institute for 25 years, while the business school operates the Dingman Center for Entrepreneurship. Both offer services to start-ups. Maryland also runs weekend "technology start-up boot camps" that draw up to 600 from outside the university who want to launch companies. It has even organized a local network of angel investors.

The university runs the oldest small-business incubator in the state and a "bioprocess scale-up facility" that develops commercial production processes, Dr. Mote explained. It also offers a state-funded consulting practice that has been replicated around the United States in which faculty help companies commercialize products. In addition to spawning a number of start-ups, many of which are based in an adjacent science park that is responsible for 6,000 jobs, the University of Maryland co-developed products ranging from power tools and telecom systems to boat sails.

Nationally, the University of Maryland collaborates with several federal laboratories in energy, life sciences, aerospace, and national security and receives $500 million in federal research funding a year, Dr. Mote said. The National Oceanic and Atmospheric Administration is

---

[45]The Kauffman Foundation 2009 report, "Entrepreneurial Impact: The Role of MIT" details a different but also interesting account of university innovation ecosystem. Available at <http://web.mit.edu/newsoffice/images/kauffman.pdf>.

establishing a global climate-change and weather-predication center at the university research park, while the National Institute of Science and Technology is contributing funds for a new lab building on campus devoted to quantum physics.

The University of Maryland also has an extensive relationship with China. Its Institute for Global Chinese Affairs, for example, has trained 3,000 Chinese executives since 1995, while 160 Chinese executives have received one-year degrees from Maryland's Executive Master's in Public Administration program. The university also has a special "international incubator" that has helped launch 11 Chinese companies in industries such as solar energy and software. In 2002, the Chinese government and Maryland set up a joint research park near campus that now houses facilities of companies from Beijing, Shanghai, and Guangzhou.

COOPERATION IN INFORMATION AND COMMUNICATION TECHNOLOGIES

Participants in the conference noted that 21st century innovation systems are based on state-of-the-art data and telecommunications infrastructure. Chen Ying of China's Ministry of Industry and Information Technology noted in his presentation that information and communications technology (ICT) has become an increasingly important driver of economic growth. He cited a World Bank study that concluded a 10 percentage point increase in broadband penetration rates can increase economic growth by 1.3 percentage points in developing nations and by 1.2 percent in advanced nations.[46] Over the next five years, ICT is expected to create $5 trillion in new economic activity.

China's Broadband Strategy

China views broadband infrastructure as a catalyst for new growth industries such as software, logistical services, information technology outsourcing, and a wide range of digital devices. Several years ago, Mr. Chen explained, the government set a target of 30 percent annual growth

---

[46]See Christine Zhen-Wei Qiang, "Broadband Infrastructure Investment in Stimulus Packages: Relevance for Developing Countries," Global ICT Department, World Bank, 2009. This World Bank study includes Internet and broadband, in addition to the fixed and mobile phones, in an econometric analysis of growth in 120 countries between 1980 and 2006. Results show that for every 10-percentage-point increase in penetrations of broadband services, there is an increase in economic growth of 1.3 percentage points.

for its software and information services industry and for software exports to grow 28 percent a year.[47] China's electronic commerce industry, which Mr. Chen said has been growing by around 25 percent a year, also is expected to see substantial expansion.

Large-scale deployment of broadband and improved applications can help transform the entire economy by integrating industries and bringing new sources of high-value services, Mr. Chen said. ICT technologies can revitalize many existing industries, from furniture manufacturing to chemicals, he noted, and can bring greater efficiency and cost savings to companies and government. So in addition to expanding broadband infrastructure, the government is putting a high priority on optimizing the use of ICT and integrating it into "our daily lives,"

America's Broadband Strategy

The United States also views broadband as critical infrastructure that must be used in "very interesting and innovative ways" in order to stimulate economic growth, Eugene J. Huang of the Office of Science and Technology said in his presentation. The Obama Administration is focusing on the "entire ecosystem surrounding broadband and how we will use it in the future," he said.

The Recovery Act earmarked $7.2 billion in grants to stimulate broadband deployment throughout the U.S and required the Federal Communication Commission to develop a National Broadband Plan.[48] The plan's "extraordinarily ambitious" goals cover not only broadband access but also its use in everything from health care to managing household energy consumption. The plan calls for affordable access with download speeds of at least 100 Mbps to 100 million U.S. homes and affordable access to at least 1 gigabit-per-second service for key institutions such as schools, hospitals, and government buildings in every community. Another goal is that the United States should have the world's fastest and most extensive wireless networks.

---

[47]China's Eleventh Five-Year Plan (2006-2010) also calls for producing around 15 major software enterprises with sales exceeding RMB 10 billion. For a good analysis of China's information technology and communication strategy by Indian software-industry association Indian software-industry association NASSCOM, see "Tracing China's IT Software and Services Industry Evolution," whitepaper prepared by NASSCOM Research, August 2007, (<http://www.business-standard.com/general/pdf/082107_01.pdf>).

[48]See Federal Communications Commission, *Connecting America: The National Broadband Plan,* <http://www.broadband.gov/download-plan/>.

In terms of access, the Department of Commerce is using Recovery Act funds to expand broadband infrastructure, public computer centers, and sustainable adoption of broadband service, Mr. Huang said. The Department of Agriculture is spending $2.5 billion to deploy broadband in rural areas. The aim is to make it feasible for every home to connect to high-speed Internet, he said, "along the lines of what the United States did in the 1930s, when it determined it was a priority to get telecommunications distributed throughout the U.S."

These investments are part of a larger strategy to use broadband infrastructure to promote economic growth and achieve national priorities. The Recovery Act included $15.5 billion to develop and implement smart-grid technologies, for example, and $19 billion to accelerate adoption of information technology in health care. Broadband also is key to new Administration initiatives in public safety communications and improving government efficiency, transparency, and public services, Mr. Huang said.

Co-developing ICT Products in China

Advances in information and communication technologies are fundamentally transforming the process of innovation itself. They are enabling enterprises, for example, to increasingly innovate "globally in a fashion that is inclusive and connected across our borders," Mark E. Dean of IBM Research explained in his presentation. Indeed, the global innovation system has become so integrated that "innovation in isolation is not significant for a successful company or one country," he said. "Most of the challenges and opportunities facing us can only be addressed with global collaboration and innovation."

IBM Research is a good example of a truly global organization. It has a network of eight major labs employing 3,000 researchers in six nations, including a 200-engineer basic research lab in Beijing. IBM also has a 5,000-engineer software application and services development lab in China. In all, half of IBM's research and 60 percent of its 220,000 technical employees are outside the United States.

IBM's goal is to create technologies that will have a global impact. "We work hard to avoid innovation in isolation, because that will create very narrow solutions that have very narrow upside potential," Dr. Dean said. So IBM is creating a matrix that involves all research labs. "There is not a single project we have across the research division that is isolated to a single country," he said.

The company also co-develops products with other companies around the world. IBM has 10,000 partners in 350 cities in China. They include

Futong, Digital China, Kingdee, and Yucheng Technologies. IBM has 100 joint labs and technology centers with Chinese universities and offers curricula that have helped trained 860,000 Chinese students and 6,500 teachers.

One of IBM's biggest collaborations in China is in Shenyang, in the northeastern province of Liaoning. IBM, the municipal government, and Northeastern University forged a five-year, $40 million partnership to develop information and communication technology to manage systems such as water purity, energy, food safety, and integrated urban planning. Dr. Dean predicts such efforts will be replicated around the world.

COOPERATION ON MEDICAL RESEARCH

Medical research is an important area of collaboration between the United States and China. . This collaboration is driven by mutual interest in finding remedies for chronic diseases. As was America's experience as its population aged, cancer and other chronic diseases are overtaking infectious diseases in China as the top killers and as a "major health care crisis," explained Anna Barker of the National Cancer Institute in her presentation. China has 1.6 million cancer deaths a year and reported 2.2 million new cases in 2009. The crisis "will get much, much worse in the next 10 to 15 years," she said.

The United States can benefit from China's help, too, in order to accelerate the discovery of new treatments and to contain skyrocketing drug-discovery costs. Dr. Barker noted that the United States reports 565,000 cancer deaths a year and new cases are forecast to rise by at least 30 percent by 2020. [49] Annual U.S. spending on cancer treatment is expected to rise from $213 billion to $1 trillion a year. NCI sees China's large data sets on cancer cases as valuable to an empirical analysis of how variations in the human genome may relate to the development and spread of cancer. According to Dr. Barker, China is also home to a large number of microbiologists, many of whom have been trained in U.S. universities and research organizations.

Joint research by the National Cancer Institute and Chinese scientists began in the 1970s with seminal studies of cancers related to certain environments, such as near tin mines or in textile mills. Many of these studies led to worldwide regulation, Dr. Barker said. Chinese hospitals are vital U.S. partners in building new clinical trial systems, she said, not only because of the nation's large patient population but also because

---

[49]Data from American Cancer Society, 2006 Cancer Facts and Figures.

managers at China's university hospitals are often familiar with NCI, many of them having been trained there.

Cancer genomics is a particularly valuable area for Sino-U.S. collaboration that could lead to important new therapies, Dr. Barker said. Chinese researchers were among the first to identify the SARS genome. The National Cancer Institute is working with Chinese institutes on an ambitious project to sequence genomes of all cancers. It also is partnering with the Beijing Genomics Institute, the world's largest next-generation sequencing center, in brain-tumor research.

Nanotechnology, which will "touch everything we do in medicine in the next 10 years," is another area of "very strong collaboration," Dr. Barker said. Five thousand scientists at 50 Chinese universities, 20 Chinese Academy of Sciences Institutes, and 300 nano-technology enterprises focus on the field.[50] The third meeting between U.S. and Chinese medical researchers on nanotechnology will be held in fall 2010.

The National Cancer Institute wants to keep expanding its Chinese partnership. Future health care research "is going to be a very distributed enterprise," Dr. Barker predicted. "But I think it will be dominated by the U.S. and Chinese because we are making the investments."

SOME CHALLENGES TO CLOSER COOPERATION
WITH CHINA

Ambassador Wolff observed that America can learn from China's search for solutions to common challenges. For example, the United States should study China's financial support for renewable-energy projects, its approaches to carbon sequestration, and the balance between large enterprises and small- and midsized firms. "We should learn something from each other by comparing these two sets of national policies," he said.

Mr. Yang noted that Sino-U.S. collaboration has "some areas for improvement," Going forward, a number of frictions between China and the United States can be worked out, allowing collaboration on innovation to move to a deeper level.

Trade

The divergent paths in innovation policy may require the United States and China to recalibrate their trade relationship, said Anna Borg of

[50]Data from *Science* 309: 65-66, 2005.

the State Department in her presentation.[51] There needs to be a "frank discussions" she said, about a "broader constructed trade framework supported by generally accepted rules and international institutions." At the same time, there was acknowledgement from individual Chinese and American participants that U.S. technology export curbs and visa policies are also significant obstacles to closer collaboration.[52]

Intellectual Property Protection

Ms. Borg also noted that some Chinese practices hurt innovation, such as weak protection of intellectual property rights. To maintain a successful innovation environment, nations must "embrace and enforce an intellectual property system that allows innovators to reap the benefits of their ideas and reward their risk-taking," she said. "Without it there is little or no incentive for companies to produce new products or services." She said cooperation on protecting copyrights and trademarks in industries such as software, drugs, music, and fashion "will go a long way in deepening" Sino-U.S. cooperation in innovation.

Weak IPR enforcement also can make it harder for nations to fully benefit from global innovation networks, Ms. Borg suggested. "Nations that fail to protect intellectual property will find themselves cut off from these dynamic global partnerships because innovative firms will hesitate to invest in or form partnerships with countries where their intellectual property may be stolen," she warned.

Chinese officials countered that their nation has made tremendous progress is establishing laws to protect IPR rights and courts to settle disputes. Enforcement "is not only the work of the government," Mr. Wang said. "Enterprises should provide evidence of IPR infringement. With evidence, a court will make a ruling." Mr. Chen of the Ministry of

---

[51]For a review of U.S. China trade issues, see Wayne M. Morrison, "China-U.S. Trade Issues" Congressional Research Service, June 1, 2010. See also *New York Times*, September 14, 2009, "China-U.S. Trade Dispute Has Broad Implications."

[52]See, for example, the remarks by Yang Xianwu of the Chinese Ministry of Science and Technology, who noted that the United States still places some restrictions on exports of high-tech products to China. In addition, he claimed that high-level personnel from China continue to encounter unpleasant experiences in obtaining visas to the United States. Responding to a question at the conference, Dr. Anna Barker of the National Cancer Institute said that obtaining visas for Chinese counterparts was a significant barrier for the first year after the September 11, 2001, terrorist attacks, but added that this situation has significantly improved.

Industry and Information Technology also noted that the government has ordered all manufacturers of computers in China to pre-install only legal operating systems. Ninety percent of computers released by the 22 largest hardware manufactures in China have legal operating systems, he noted. "We have the hard figures to prove that, at least at the operating system level, the piracy issues have greatly improved," Mr. Chen said.

'Indigenous Innovation'

Ambassador Wolff challenged China's focus on "indigenous innovation," especially at a time when knowledge flows with increasing ease and speed throughout the world. "In this globalized world, there is no indigenous innovation," he contended. The question is whether "on balance, these policies are helpful or harmful to China."

Ms. Borg also said that policies favoring domestic innovation could backfire. Some of the greatest benefits of innovation come from adopting innovations of others. Investment barriers or domestic intellectual-property requirements "will ultimately be self-defeating," she warned. "In the short run, China's entire economy will be less competitive when it is denied access to the full range of innovative products available in the global market."

China's own creative industries will be stifled if they are denied exposure to international competition and new technology, she said. Requirements that government agencies buy locally developed technology also "constitute a step toward import substitution" and "invite retaliation," Ms. Borg said. As they seek to boost their investments abroad, Chinese companies also will benefit from a transparent regulatory and legal environment, she said.

Despite the growing emphasis on indigenous innovation, China still attaches great importance to international cooperation in science and technology, Mr. Yang of the Ministry of Science and Technology said. He acknowledged that China's innovation system "cannot be separated from the rest of the world."

He noted that China has signed science and technology cooperation relationships with 152 nations and regions, sent science diplomats to 45 nations, and has joined 350 different international science and academic organizations, in which 265 Chinese scientists hold posts. China has participated in the Human Genome Project and European Galileo Program, which is developing a satellite for geo-positioning systems. China's main objection is to "absorb innovation" from the outside and adapt it to serve "Chinese conditions."

Few relationships have been more important than that with the United States, Mr. Yang said. He noted that the United States and China have signed some 50 cooperation agreements in fields such as agriculture, energy, and medicine involving nearly every Chinese government agency.

The most recent agreement, to establish the Sino-U.S. Joint Research Center for Clean Energy, is one of the most significant. For the first time, each nation will contribute an equal amount of money and assign scientists to an independently managed research center focusing on clean water, clean air, and other areas. "This represents an historic point," Mr. Wang said. "In the past, cooperation mainly focused on exchanges of personnel. This is the first time both governments donated directly to a joint program."

Equal Treatment for U.S. Firms in China's Markets

Another U.S. complaint is that foreign companies collaborating in R&D in China aren't treated as equals when it comes to the domestic market.[53] Although IBM works with the government, "we're not viewed as a Chinese company, which can be a constraint in many ways," said IBM's Mark Dean. "We would like to be viewed as an equal partner, because we believe our investments will be on par with those of Chinese companies."

There also was a sense by some on the American side that the United States often gives more than it benefits from R&D partnerships with the Chinese. If cooperation is to work, "it must be based on an equal exchange," Michael Borrus of X/Seed Capital remarked. "Each side must give as well as get."

To move towards greater reciprocity, the key to success may lie in *creative incrementalism*. As suggested by Michael Borrus in his concluding remarks: "We need to try some things together, demonstrate mutual gain, and then turn those smaller-scale collaborations into larger collaborations."

---

[53]The U.S. Chamber of Commerce has called the regulatory environment in China increasingly difficult for foreign companies citing government procurement rules that favor local companies, a postal law that excludes foreign suppliers such as FedEx Corp. and curbs on rare-earth exports. See *New York Times*, January 18, 2011, "U.S. Shifts Focus to Press China for Market Access."

BUILDING ENDURING FOUNDATIONS

For all of the philosophical differences voiced in the symposium, the common theme was on the strong foundations available on which to build U.S.-China cooperation in science and innovation. Individual participants from both the United States and China voiced their support for the basic elements of a globally connected innovation system, which include developing strong commitments to open scientific and applied research, education, spurring corporate R&D investment, and growing international partnerships, especially in areas of compelling mutual interest such as medicine and energy.

The mission of universities also is expanding in both countries to pay more attention to commercial applications. "Universities should remain focused on discovery of new scientific knowledge, new technologies, and new processes," noted Dr. Vest. "But I think they are going to be increasingly use-inspired. People are simultaneously exploring the unknown, but with a broad end-goal in mind," he said.

There also was evidence of some convergence in philosophy. China is trying to transform an innovation system dominated by state institutions into one driven more enterprises and the market. "We learned from advanced countries," Mr. Wang said. The United States, by contrast, is searching for a more effective and impactful role for public policy and federal agencies. Nations with state-led innovation systems "are all trying to work their way to the bottom," observed the University of Maryland's Mote, "while the United States is trying to work its way to the top."

Which mix of innovation policies and investments proves most effective in tackling enormous global challenges such as climate change, energy, and medical care for aging populations remains to be seen. As we see in the proceedings, summarized in the next chapter, the participants in this workshop highlighted a variety of areas where cooperation between China and the United States can help address these global challenges.

II
PROCEEDINGS

WELCOME

Charles Wessner
The National Academies

Dr. Wessner welcomed the guests from China and the United States assembled in the National Academy of Sciences to discuss building bilateral cooperation in science, technology, and innovation. Within these walls "we often talk about science and sometimes about technology," Dr. Wessner noted. "We are learning to talk more about innovation." He added that the National Academies Board on Science, Technology, and Economic Policy (STEP has a "somewhat unique mission" to integrate the diverse elements of science, technology, and economics in order to generate policy recommendations for the U.S. government. Many of these policy recommendations are adopted by Congress and the Administration.

Dr. Wessner noted that STEP has underway a comparative assessment of national innovation policies. This program is studying innovation policies of major nations, such as Japan, India, leading European nations and regions—and China.[1]

A real-world understanding of other nations' practices and experiences is important for U.S. policymakers, Dr. Wessner said. "One of the things we struggle with here in the United States is that some

---

[1]For examples of previous comparative studies, see National Research Council, *Innovative Flanders: Innovation Policies for the 21st Century,* Charles W. Wessner, ed., Washington, DC: The National Academies Press, 2008. Materials from the September 24-25, 2007, STEP conference "The Dragon and the Elephant: Understanding the Development of Innovation Capacity in China and India" may be found at *<http://sites.nationalacademies.org/PGA/step/PGA_046383>*.

people seem to understand the world better in theory than in practice. These people often have a powerful influence," he stated. In comparison, STEP is looking less in theory and more in fact about what the rest of the world is doing.

The National Academies also is interested in expanding mutual cooperation. "With almost everything we need to do to make the 21st century a more prosperous century, safer century, and more environmentally friendly century, China and the United States must work together," he said.

Dr. Wessner noted that this conference was organized with the assistance of Cisco Systems Inc. He also thanked the program's other sponsors. They include International Business Machine, Intel, the Palo Alto Research Center, Sandia National Laboratories, the Office of Naval Research, the Defense Advanced Research Projects Agency, the National Institute of Standards and Technology, and the National Science Foundation. He offered special thanks to Patrick Keating, Cisco's director of worldwide leadership education, "whose leadership and common sense have done a great deal to make this program possible."

Dr. Wessner then introduced the keynote speaker, Ambassador Alan William Wolff, a former U.S. trade ambassador and chairman of the Committee on Comparative National Innovation Policies. Ambassador Wolff also is a research professor at the Monterrey Institute of International Studies and counsel at the Washington law firm Dewey & LeBoeuf.

OPENING REMARKS

Alan Wm. Wolff
Dewey & LeBoeuf LLP

Ambassador Wolff welcomed the delegation from China and American participants on behalf of the Science, Technology and Economic Policy Board of the National Academies.

The United States and China have cooperated in science for at least 70 years, noted Ambassador Wolff, a prominent trade attorney who chairs the STEP Board's Committee on Comparative National Innovation Policies. Prior to World War II, he noted, the United States allocated precious air cargo space to ferrying scientific instruments, materials, and current treatises over the Himalayas from India to Chongqing so that Chinese scientists in exile could continue their work during Japan's occupation of China.

The modern history of Sino-U.S. science and technology cooperation began on June 5, 1965, in the same National Academy of Sciences building that was the site of this symposium. On that date, the Academy decided to create a committee to foster academic communication and exchanges with China.[1] The Council stated: "We hopefully believe the U.S. scientific community can contribute to a lessening of tensions between peoples and nations by endeavoring to create the basis for scientific discourse between Chinese and American scientists."

That move was important, given the historical context. "You have to remember what great difficulties there had been in the recent past," Ambassador Wolff recalled. American and Chinese troops had fought in

---

[1]For historical background on Sino-U.S. cooperation in the 1960s and 1970s, see Kathlin Smith, *The Role of Scientists in Normalizing U.S.-China Relations: 1965-1979*, Council on Library and Information Resources (<*http://china-us.uoregon.edu/pdf/Smith's%20NYAS%20article.pdf*>).

37

the Korean Peninsula, and anti-People's Republic of China attitudes among some in Congress were strong. "So it was a difficult period for this academy to begin to reach out to colleagues in the Chinese sciences to build bridges," he said.

The Cultural Revolution soon intervened, however, and it wasn't until the 1970s that small numbers of U.S. scientists began to visit again. There was strong American interest in China's studies of botany and seismology, areas in which China was advanced, Ambassador Wolff noted. The Chinese scientific community, meanwhile, was interested in topics related to the nation's industrial and agricultural priorities, such as computer science, petrochemical engineering, mineral extraction, telecommunications, mechanized agriculture, and industrial automation.

Exchanges resumed in earnest after the Nixon-Zhou Enlai 1972 Shanghai Communiqué. In 1978, Deng Xiaoping suggested there was potential for expanding bilateral exchanges. Ambassador Wolff, who served as U.S. deputy special representative for trade negotiations at the time, noted that the first high-level science delegation to China that year was led by a colleague of his, Frank Press, then President Jimmy Carter's science advisor and later president of the National Academies.[2] That trip, he said, "provided the foundation for the formal bilateral understandings to foster science and technology cooperation that followed." Soon afterward, China's Ministry of Science and Technology and America's National Science Foundation resumed formal cooperation.

The Sino-U.S. partnership in science and technology played an important role in helping China's scientific community recover "from the dislocations of the Cultural Revolution," Ambassador Wolff said. Meanwhile, "American universities were and are an enormous source of education for Chinese students. Investment in China by American and other foreign corporations was and is an important source of technology for China."

Now, the United States is starting to benefit. "We may be on the threshold of some reverse flow of investment, from China to the United States, and China's graduate students enrich the research environment of American universities," Ambassador Wolff observed. "The fruits of major research activity that will take place in China will be available to other countries as well." One recent sign of this trend, he noted, is that Applied Materials Corp.'s chief technology officer is moving to China to improve production of solar-panel equipment.

---

[2]Frank Press served as presidential science advisor from 1977 through 1980 and as president of the National Academies from 1981 to 1993.

Perhaps just as important as the flow of scientific knowledge is the exchange of ideas on science and technology policy, Ambassador Wolff said. He recalled a comment made at a 2006 conference in Beijing by Richard C. Atkinson, the former director of the National Science Foundation. Mr. Atkinson explained that in the 1970s "there was very little economic theory or data about investments in R&D and economic development [to make] the case to the Congress for federal support of research." The NSF, therefore, initiated a study exploring that link. Decades later, Dr. Atkinson noted, a report by the President's Council of Economic Advisors concluded that half of the growth in the American economy in the previous 40 years had been due to investments in research and development.[3] "The private sector is a major driver of R&D, but federally funded research at universities plays a key role," Dr. Atkinson said in his 2006 speech.

Atkinson's message to his Chinese counterparts was that he believed government funding of university research "was a core need of scientific progress and innovation in this country," Ambassador Wolff explained. In the United States, supporters of science and technology are in "a battle right now to make sure that government funding of basic research and development is sufficient," he noted. "Many people around this room and in the American scientific and technology community are currently trying to get through Congress a very important level of funding for university research and other research in this country. So the struggle continues and hasn't changed that much in 40 years."

The scientific communities of China and the United States can benefit from sharing views on best practices for national policies, he said. "Today is another step in that process of mutual exchange and—I trust—mutual benefit."

Ambassador Wolff then explained America's innovation system. A 2006 National Academies publication defined the National Innovation System, a term popularized by President Richard Nixon, as "a network of institutions in the public and private sectors, whose activities and interactions initiate, develop, modify, and commercialize new technologies." This National Innovation System, the publication explained, involves flows of knowledge among complex, inter-linked, and overlapping "innovation eco-systems" at universities, government research laboratories, large and small businesses, and other organizations.[4]

---

[3]Council of Economic Advisors, *Economic Report to the President*, 1995.
[4]This definition is cited in National Research Council, *India's Changing Innovation System: Achievements, Challenges, and Opportunities for*

China also has an immensely rich history of innovation. "For many centuries, if not millennia, China led the West in innovation," Ambassador Wolff observed. "Not only were remarkable things invented, but they also were put into circulation for practical use." In the West, schoolchildren learn that the world is indebted to China for inventing porcelain in the 7th century AD; gunpowder, fireworks, and rockets in the 4th century AD; paper and tea in the second century BC; kites in the 5th century BC, and silk in 3600 BC. "But I don't think many outside China know that the invention of noodles dates back at least 4,000 years," he said.

Ambassador Wolff then presented an extensive sampling of other, less-heralded Chinese inventions:

Magnetic Compass	**200 BC**
Movable Type	**1050 AD**
Wrought Iron	**5th Century BC**
Blast Furnace	**250 BC**
Paper Money	**700 AD**
Paddle Wheel Boats	**650 AD**
Metal Bells	**200 BC**
Fork (preceded chopsticks)	**2400 BC**
Lacquer Ware	**5000 BC**
Stone Plowshares	**3500 BC**
Toxic Gas for War	**400 BC**
Use of Chromium (for weapon tips; first used in West around 1797)	**210 BC**
Golf	**1000 AD**
Crossbow	**200 BC**
Use of Vitamin-Rich Foods (as disease treatment)	**200 BC**
Diagnosis of Diabetes	**200 BC**
Dietary Treatment of Diabetes	**650 AD**
Isolation of Hormones (used for medical treatments)	**1110 AD**
Fishing Reel	**4th Century BC**

Cooperation, Charles W. Wessner, ed., Washington, DC: The National Academies Press, 2007.

Manned Flight With Kites (1891 in Europe)	**6th Century AD**
Standardized Lumber Dimensions	**1100 AD**
Natural Gas Use for Heat and Light	**4th Century BC**
Negative Numbers (also in Greece, but not used widely in Europe until 1550)	**3rd Century AD**
Pinhole Camera (a century before discovery by Aristotle)	**450 BC**
Raised Relief Maps	**3rd Century BC**
Rotary Cooling Fan (first used in West in 16th Century)	**200 BC**
Seismometer	**132 AD**
Steel	**2nd Century BC**
Iodine Treatment for Goiter (1860 in France)	**7th Century AD**
Chain Suspension Bridge	**15th Century**
Toilet Paper	**589 AD**
Tune Bells	**8th Century BC**
Underwater Salvage	**1065 AD**

Only in more modern times has the technology flow begun to reverse from West to East, Ambassador Wolff noted. The flow began with innovations like the windmill from the Middle East and telescope from Europe, and continued with a "cascade of inventions borne of the industrial and information technology revolutions," he said.

For the past decade, Chinese leaders have stressed that innovation is vital to the nation's future. In 1999, Ambassador Wolff noted, General Secretary Jiang Zemin said in a speech at a conference on innovation: "In today's world, the core of each country's competitive strength is intellectual innovation, technological innovation and high-tech industrialization.[5]

Six years later, Party General Secretary Hu Jintao introduced a new objective, when he said that the government should "give priority to indigenous innovation" in science and technology work. Mr. Hu also said the country should "increase core competitiveness and strive to make science and technology innovation with Chinese characteristics a

---

[5]Jiang Zemin, General Secretary of the Communist Party of China Central Committee, keynote speech to the National Technological Innovation Conference, August 23, 1999.

reality."[6] He also said the government must "create a policy environment beneficial to technological innovation."

The United States also has grown increasingly concerned about advancing innovation. Ambassador Wolff cited the landmark 2007 report titled *Rising Above the Gathering Storm,*[7] produced by a committee of the National Academies. The report said the government should design science and technology policy to:

> "...ensure that the United States is the premier place in the world to innovate; invest in downstream activities such as manufacturing and marketing; and create high-paying jobs based on innovation by such actions as modernizing the patent system, realigning tax policies to encourage innovation, and ensuring affordable broadband access."

There are many similarities between the innovation goals and policies of the United States and China, Ambassador Wolff said. "Each wishes to enhance the prospects for successfully initiating, developing, modifying, and commercializing new technologies." Both countries also "want a substantial part of all stages of the innovation system to be located within their own national boundaries." This does not necessarily mean each product must be developed locally, "but at least a healthy share of the spectrum for products in general" so that a large number of high-quality jobs are created to bring economic benefits.

The focus is different, however. Chinese leaders seem to be more concerned with the "front end of this process, initiating and developing new technologies," he said. The United States is more concerned with the "back end, the commercialization of new technologies."

There also are some similarities and differences when it comes to policy. "We should learn something from each other by comparing these two sets of national policies," he said. Both the United States and China recognize the need to support science, technology, engineering, and math education from the primary school level to advanced degrees, he said. They also both "recognize the importance of supporting university

---

[6]Speech by Hu Jintao, General-Secretary of the CPC Central Committee, November 27, 2005.
[7]National Academy of Sciences/National Academy of Engineering/Institute of Medicine, *Rising Above the Gathering Strom: Energizing and Employing America for a Brighter Future,* Washington, DC: The National Academies Press, 2007.

research, which has not been true of every leading trading country," he said. The United States and China both support research parks, regard protection of intellectual property as important, and "see a global interest and a national interest in creating renewable energy technologies and associated equipment industries and utilities," he said.

Policy differences between China and the United States, however, "require examination," Ambassador Wolff said. One is U.S. immigration policy. As a result of tougher American immigration and work-visa policies, he said, the United States is having a harder time retaining highly trained, foreign-born talent with advanced degrees in science and technology.

Another issue is U.S. defense spending. Early investment by the military was instrumental in the commercial success of integrated circuits, the Internet, large aircraft, and GPS navigation systems, Ambassador Wolff pointed out. Government demand, which does not move as quickly as private commercial demand, can also be a drag on the pace of technological evolution. Today, for the bulk of products that are not exclusively used by the military, commercial demand is a powerful catalyst for development. The government can help launch technology, not guarantee widespread adoption. "The involvement of government is something like booster rockets for the Space Shuttle, which must not remain attached after initial thrust," Ambassador Wolff said. If attached, "they would make getting into orbit impossible."

The United States is closely watching a number of developing policies in China. For example, it is interested in whether China's heavy investments in infrastructure will present market opportunities for foreign companies. "There will be global commercial benefits for China from the creation of its high-speed rail industry and photo-voltaic cells, among other industrial policy programs," he said.

The United States also is watching China's policies with respect to intellectual property and whether it is adopting national rather than international industrial standards. The evolution of China's National Indigenous Innovation Policy is another major issue. The question is whether, "on balance, these policies are helpful or harmful to China," Ambassador Wolff said.

The two nations are interested in the policy tools each other are using to promote innovation. One question, for example, is how China's financial support for renewable-energy projects compares with those of others. Another topic of mutual interest is the role venture capital will play in both countries compared to other sources of capital. The related issues include "what is needed, at what stage, and is it forthcoming?" Ambassador Wolff observed. Other topics include the optimum role of large corporations, including state-owned enterprises in China, as compared with small and medium enterprises. Yet another is future role of foreign direct investment in promoting innovation. "Each of these topics and many others will emerge from our discussions and invite further exploration," he said.

Conferences such as this one help stimulate thinking on ways to improve both nations' science and technology policies, Ambassador Wolff said. Among the fundamental questions: "Which policies promote innovation and which may retard or distort the process?" he said. "Where are there areas for future cooperation and collaboration? Where can we find areas where working together would have the potential for creating a major benefit for other countries as well – such as finding solutions to the challenges of carbon sequestration, cheap energy-efficient bio-fuels and batteries?"

In his own view, Ambassador Wolff said, government support is "vitally important for progress in science, technology, and innovation." However, government should play a supporting role, "like that of a proud mother or father watching a high school or college graduate." Government direction that is "warranted and truly useful after that graduation is limited," he said. "As much harm as good can come from such interventions. That is our national experience and our bias. So interventions must be very careful so as not to be counterproductive."

The market drives perhaps as much as 85 percent of innovation, Ambassador Wolff said. "If we were clever enough to figure out where the market was headed, we would all be billionaires."

Ambassador Wolff also challenged the notion of indigenous innovation as a useful path. "In this globalized world, there is no indigenous innovation," he said. Before the rise of fast and easy international communication, local innovation was more common. Ambassador Wolff noted that in his books on science and innovation in China, the eminent historian Joseph Needham[8] credited China with invention of the stirrup, which allowed warriors to stay in the saddle at a full gallop. "But the Hittites, my wife tells me, invented the stirrup about two millennia earlier," Ambassador Wolff said. "What the Hittites and the Chinese inventors lacked was access to Internet cafés to cross-fertilize their innovations for their mutual benefit."

---

[8]Joseph Needham (1900-1995) edited a series of volumes on *Science and Innovation in China* published by Cambridge University Press. The first of 27 volumes was published in 1954. The project continues under the Needham Research Institute.

New technologies will make such cross-fertilization even easier. Ambassador Wolff noted that Cisco has a great video-conferencing system that makes individuals at a meeting feel like they are in the same room as those across the table from them, even though they actually are on different continents. The next step will be use of holograms, which will enable people to feel the presence of others located far way and to conduct "truly virtual meetings," he said. That will save money not only on tea and pastries but also airfare, he noted.

Science, technology, and economic policy usually catch up with progress, Ambassador Wolff observed. "But policy often lags invention. It does not very often precede it." He noted that the integrated circuit and Internet "were great enablers of this new world, but the applications that add the next very large layers of value are the products of individual and private corporate achievements."

Each country and industry needs to find best practices to support innovation and inform each other of these findings, Ambassador Wolff said. "This is a secret recipe for progress. When we discover it and share it, we enrich the world." He said the United States and China have much to learn from each other and added that he hopes the two nations will follow up on this conference with another conference in China so that the dialogue can continue.

OPENING REMARKS

Ren Weimin
National Development and Reform Commission

Mr. Ren began by noting he was "very impressed" by Ambassador Wolff's list of Chinese inventions. "I don't think Chinese people have thought very deeply about our innovations. We had always assumed we've got the Four Innovations," he said. "Over our long history, many things we considered to be normal for the survival and everyday living of our ancestors may well have been innovations. So I agree on this way of looking at our history and culture."

China and the United States should focus on cultural exchange so that the two nations can better understand each other, Mr. Ren said. "Even after several thousand years of history, each nation has its own unique mindset and thinking mode," he said. "If we get to know each other from this perspective, we can better handle our bilateral relations."

Mr. Ren expressed the gratitude of the Chinese delegation for being invited to visit the United States. "The international financial crisis has not been fully alleviated, and the world economy revived, during this trip," he noted. "We have had an exchange of views on issues like energy innovation. We have benefitted a lot." The United States remains the most advanced country in terms of technological exchange, Mr. Ren said. "This program has broadened our horizon, as well as our understanding of many aspects of the U.S."

Even though the most difficult period of the recent global recession appears to be over, "the legacy of this financial crisis has not left us yet," Mr. Ren said. "An entire revival of the global economy needs all countries and industries to better have better understanding and co-operation. Even though there may be some disparity of opinions, we have reached common ground on many issues."

The United States and China have "tremendous potential" to cooperate in high-technology industries because they have complementary strengths, Mr. Ren said. "I hope we can deepen the Sino-U.S. friendship and that our cooperation with be continuously fruitful," he said. "I hope this conference today will be very great success."

BUILDING GLOBAL PARTNERSHIPS:
OPPORTUNITIES IN U.S.-CHINA COOPERATION

Introduction

Dr. Wessner expressed delight at having Anna Borg to speak at the symposium in place of Under Secretary of State for Economic, Business, and Agricultural Affairs Robert D. Hormats. Under Secretary Hormats, ironically, had been called away to travel to China. Ms. Borg, assistant secretary to the Economics Bureau at the State Department, has a "distinguished career in the foreign service," Dr. Wessner said. He mentioned that when he tried to reach Ms. Borg the previous day to confirm she could speak on short notice, he asked her office to contact her on her cell phone. "Their answer was that the White House doesn't like people being called when they visit the White House," Dr. Wessner said. "I think that gives you some indication of the role Assistant Secretary Borg has."

Anna Borg
U.S. Department of State

Ms. Borg said she was happy to be invited to speak at the symposium. Ms. Borg noted that Under Secretary Hormats has spent a "tremendous amount of time recently in China" and that many State Department officials were going to China for an upcoming U.S.-China Strategic and Economic Dialogues.[1] "This is a prelude to that, which occurs in just a few days."

---

[1] The U.S.-China Strategic and Economic Dialogue is a series of high-level bilateral meetings established by President Barack Obama and Chinese President Hu Jintao in April 2009 to discuss a broad range of issues between the two nations.

Ms. Borg added that it was "very enjoyable to hear about all of the different innovations that have come out of China and to hear from our Chinese colleague about some of the thoughts he has in regard to innovation."

During the day's discussions, "we realize that not only do we innovate, but we innovate very often when we look at challenges that we face," Ms. Borg said. "Climate change, energy shortages, disease epidemics, famine, and terrorism are just a few that come to mind. Innovation—the development of new ideas and products—is necessary to offer previously unthought-of solutions to these hurdles."

As the world's largest economy and the world's fastest-growing economy respectively, the United States and China "share an opportunity and an obligation to work together to promote and protect innovation," Ms. Borg said. "Cooperation between the governments of the United States and China as well as its citizens and businesses are imperative to solve the problems of today and tomorrow."

Three key areas where the United States and China should work together are "creating an environment that favors innovation; maintaining an open, rules-based trade system; and advancing efficient and sustainable energy policies," Ms. Borg said.

To create an environment that fosters innovation, "countries need to get a range of policies right," Ms. Borg said. These policies include education, research-and-development funding, good governance, transparent regulatory policies, open and competitive markets, and "policies that allow companies to succeed and sometimes fail." She said nations "must also embrace and enforce an intellectual property system that allows innovators to reap the benefits of their ideas and rewards risk-taking."

Intellectual property promotes innovation, Ms. Borg explained. "Without it there is little or no incentive for companies to produce new products or services." Copyrights, trademarks, patents, and trade secrets that protect creativity, entrepreneurship, and innovation are key drivers of domestic and global economic growth, she observed. "Therefore, the theft of IP continues to be a concern. Emerging nations like China need to rigorously protect intellectual property rights for their own companies and for foreign companies." The latter, she added, should be treated "fairly, just as (governments) would want their businesses treated abroad."

The U.S. government would like to work with China's government "to ensure that the rights of all intellectual-property holders, such as in the software, pharmaceutical, music, and fashion industries, are well-protected and that laws are consistently enforced," Ms Borg said. "We hope that China shares these objectives and will work with us in fostering an innovation climate." She said that such support "will go a long way in deepening" the two nations' cooperation in innovation.

An open attitude toward innovation is essential in today's global economy, Ms. Borg said. Companies are forming innovation networks that include other firms, customers, suppliers, universities, and government institutions around the world. Their products incorporate technologies from a number of countries and companies. "Rarely are such complex products based solely on the intellectual property of a single business or a single nation," she said. "Nations that fail to protect intellectual property will find themselves cut off from these dynamic global partnerships because innovative firms will hesitate to invest in or form partnerships with countries where their intellectual property may be stolen."

China and the United States also must work together to promote open trade in order to promote innovation, Ms. Borg said. Both nations have a stake in an "open and rules-based global trading system," she said. "There is need for frank discussions between our two countries about broader constructive trade frameworks supported by generally accepted rules and international institutions."

Ms. Borg said she recognizes China has made development of local creative industries a top priority. Some of the greatest benefits of innovation, however, come from adopting and adapting the innovations of others. "The imposition of barriers in the form of performance, investment, or intellectual property requirements to achieve this goal will ultimately be self-defeating," she warned. "In the short run, China's entire economy will be less competitive when it is denied access to the full range of innovative products available in the global market. In the long run, China's own creative industries will be stifled when they are denied the benefit—and it is a benefit—of international competition and exposure to new technology."

The Chinese government's "indigenous innovation" policies are of particular concern to the United States, Ms. Borg said. "We applaud China's goal to become an innovative society by the year 2020," she said. But the U.S. government is concerned about the recently introduced indigenous innovation accreditation system and its requirement that government purchases be linked to products developed domestically and Chinese-owned intellectual property. The United States suspects such policies "constitute a step toward import substitution," Ms. Borg said.

The United States wants entrepreneurs and researchers in places such as Silicon Valley and Tianjin to work and benefit together. "That will not happen if there are restrictive and nationally focused procurement, standard-setting, or licensing policies," Ms. Borg said. "Protectionist policies are unsustainable because they restrict competition and invite retaliation."

China and the United States should work together to identify best practices that encourage innovation, Ms. Borg said. Making it easy for American companies to operate in China and for Chinese companies to operate in the United States is essential. Foreign investment into the United States is becoming "critical to our economic growth and job creation," she explained. "As China seeks to increase investment abroad, we want to work together to ensure a transparent environment consistent with our regulations and laws."

She pointed out that China also benefits from foreign investment. "The United States seeks fair and equitable treatment for our investors abroad," Ms. Borg said. "That is the impetus behind our many dialogues and exchanges." These discussions are conducted within the Strategic and Economic Dialogue, the Joint Commission on Commerce and Trade, and the U.S.-China Ten-Year Framework for Energy and Environment Cooperation.

The Ten-Year Framework illustrates how the United States and China can collaborate to advance technological innovation in the energy sector, Ms. Borg said. It provides a forum to promote adoption of highly efficient, clean-energy technology and sustainable use of natural resources. "As the two largest energy consumers and greenhouse gases emitters, both the United States and China have a critical interest in adapting to and mitigating the effects of climate change," she said.

The eco-partnerships that have emerged through the energy framework and other bilateral dialogues, Mr. Borg noted, have paired U.S. and Chinese cities, research institutes, and businesses to work on issues such as clean air and water, natural resource conservation, electric vehicles, and renewable energy. "Successfully meeting the clean-energy and climate challenge will help anchor U.S.-China relations in the years ahead and demonstrate to the world that our two countries can work together to effectively address global issues," she said.

In sum, Ms. Borg said, "the United States and China are, in every sense, building a global partnership." There still are many areas where the two nations "must form stronger collaborations and come to agreement," she said. "Nevertheless, I remain confident that, over time, we will continue to expand technological cooperation and collaborative innovation between our two countries."

PANEL I
BUILDING THE NEW ENERGY ECONOMY

Moderator:
Michael Borrus
X/Seed Capital Management

One of the many global issues that should unite China and the United States is the "desperate need to diversify both economies," said Mr. Borrus, the founding general partner of early-stage investment firm X/Seed Capital. Both economies must be weaned of their "dependence on sources of energy that produce massive quantities of greenhouse gases and deplete scare resources."

Leaders of both nations have embraced the goal of rebuilding economies based on renewable energy sources, Mr. Borrus noted. But they face daunting challenges, such as political pressure from powerful entrenched interests like domestic coal and oil producers.

Other challenges, he said, "come from the very different ways our two nations interact to influence global competition." A good example is in the solar energy industry. Most innovations in solar come from the U. S., he noted. But "virtually all of the manufacturing will end up in China." Over-capacity stemming from over-investment in China has caused world solar-panel prices to plummet. That has "effectively destroyed the economic case for investment in domestic U.S. solar production capacity," he said, "a situation that is not sustainable, in my opinion."

Only a combination of continued innovation and cooperation is likely to solve such problems, Mr. Borrus said. He noted that the first speaker addressing these challenges and opportunities for cooperation is Ren Weimin of the National Development and Reform Commission. He had been introduced previously. The other speaker, Energy Under Secretary Kristina Johnson, "embodies both innovation and entrepreneurship, as well as academic excellence," Mr. Borrus said. Prior to her current appointment, Dr. Johnson had been dean of Duke University's Pratt School of Engineering and provost and senior vice-president for academic affairs at Johns Hopkins University. She also has helped found two companies, performed pioneering research in optoelectronics, and was awarded the prestigious John Fritz medal.[1] Previous recipients of the prize, Mr. Borrus noted, include Alexander Graham Bell, Thomas Edison, and Orville Wright.

New Renewable Energy Initiatives in the United States

Kristina M. Johnson
U.S. Department of Energy

China and the United States share many strategic energy interests, Energy Under Secretary Johnson said. Both nations are among the world's top energy producers and consumers and emitters of carbon dioxide. "By working together we can leverage our comparative advantages in innovation and address this global climate challenge," she said.

Controlling greenhouse gas emissions is a major responsibility of both nations. Together, the United States and China emit 40 percent of the world's CO2. Europe accounts for around 20 percent. "There is no way to tag the molecules under the Chinese flag or the American flag. We own them all," Dr. Johnson noted. "So we will have to work together to figure out a way to mitigate the impact they have and will have in the future on our environment."

That impact already is becoming visible everywhere. In her home town in Colorado, Dr. Johnson said, an infestation of pine beetles has caused widespread deforestation. In a nearby state, half of glaciers have been lost in the last 100 years. "I am sure there are places that are familiar to you where you have seen the effects of climate change over the past several decades," she said.

The great news is that Americans and Chinese "both come from very innovative cultures and we know how to address challenging problems," Dr. Johnson said. "As an engineer, every problem is an opportunity to innovate. So I look forward to working together."

Dr. Johnson presented an overview of the Obama Administration's agenda to expand the clean-energy economy in the United States. The goals are to secure America's energy future, reduce greenhouse-gas

---

[1]The John Fritz Medal was established in 1902 in honor of steel magnate John Fritz and is awarded by the American Association of Engineering Societies (AAES) for important achievements in science or industry. Dr. Johnson won in 2008.

emissions by 83 percent by 2050, demonstrate science and engineering leadership, and collaborate with the global community to clean up toxic waste that is a legacy of the Cold War by 2015.

Some 70 percent of U.S. electricity is derived from fossil fuels. De-carbonizing U.S. energy generation is a challenge, but one that is "fairly straightforward," Dr. Johnson said. The United States plans to expand commercial nuclear power generation, to install carbon-capture and storage technologies in coal-fired plants, and to increase renewable energy. "We will see a fundamental shift in how we meet our electricity demand from primary sources converted to electricity," she said.

The greatest challenge is de-carbonizing the transportation sector, Dr. Johnson said, which accounts for 29 percent of U.S. energy use and puts around 2 gigatons of CO2 into the atmosphere. Of the 28 quads[2] of energy consumed by transportation, 95 percent is from petroleum.

Light-duty vehicles are responsible for about 60 percent of carbon emissions, Dr. Johnson noted. Medium to heavy trucks and buses consume another 20 percent, air transportation 14 percent, and the balance by shipping and rail. Trucking is clearly less energy-efficient per ton of goods shipped than rail. To de-carbonize commercial transportation, she said, the United States must look at advanced bio-fuels, jet diesel, and fuel cells for electric vehicles, as well as shifting more freight from trucks to rail.

Heat and electricity in buildings consumes about 39 percent of energy and releases about 2 gigatons of C02. Industry contributes a similar amount. Another priority, therefore, is to make buildings more energy-efficient. Industries of all sizes consume 32 percent.[3]

Despite the difficult economic environment in the United States, the Administration is substantially increasing investment in clean energy, Dr. Johnson explained. Under the American Recovery and Reinvestment Act, which she described as "historic legislation," $36.7 billion has been allocated to the DoE alone for this purpose—more than twice the agency's normal annual budget—and $80 billion across the federal government. These public funds leveraged $150 billion in private investment in clean-energy projects.

The new funding also realigned DoE priorities. Under the agency's base 2009 budget of $16.6 billion, for example, 41 percent of funds were devoted to R&D and 18 percent to deploying energy technologies, Dr.

---

[2] A quad refers to 1 quadrillion BTUs of energy, the equivalent of 8 billion U.S. gallons of gasoline or 293 billion kilowatt hours of electricity.
[3] Data from "State Energy Consumption Estimates: 1960 through 2007," Tables 8-12, Energy Information Administration, August 2009.

Johnson explained. By contrast, 75 percent of the DoE's Recovery Act funds—or $27.5 billion—are earmarked for deployment while 8 percent is earmarked for R&D. Other funds went to loan guarantees and other opportunities to grow the clean-energy economy.

The DoE has focused on what Dr. Johnson described as the "really difficult part of the clean-energy economy"—de-carbonizing transportation. It invested $3.4 billion to develop next-generation vehicles and fueling infrastructure. Those funds are in addition to $8.4 billion extended so far under the Advanced Technology Vehicle Manufacturing Loan Program,[4] which is outside the Recovery Act. "The projects aim to transform our transportation sector by creating competition among electrification of the fleet, hydrogen fuel cells, and compressed natural gas," she said, "as well as pushing the industry to create a pathway toward bio-fuels and more efficient commercial combustion engines."

Over the next several years, three new electric-vehicle plants will be built, Dr. Johnson noted, the first ever built in the United States. Thirty new battery and electric-vehicle component manufacturing plants also will become fully operational.

To discover new sources of fuel, the Recovery Act deployed $600 million to build 19 pilot, demonstration, and commercial-scale bio-refineries. Currently, these facilities are for ethanol, Dr. Johnson explained. In the future, the focus will shift to third- and fourth-generation fuels that involve direct conversion of sunlight into fuel.

In addition to the bio-refineries, the DoE "has tried to be innovative in the way we do innovation," Dr. Johnson said. It has launched programs such as Energy Frontier Research Centers, the Advanced Research Project Agency, and hubs.[5]

To understand how these initiatives fit together, Dr. Johnson said, it helps to understand the historical background of America's research-and-development ethic. Today's U.S. innovation system was influenced 60 years ago by Vannevar Bush, who Dr. Johnson described as a "brilliant

---

[4]The Advanced Technology Vehicle Loan program is administered by the Department of Energy. First funding of grants, loans, and other incentives to makers of automobiles and auto parts to support development and manufacturing of advanced vehicles was provided under Section 136 of the Energy Independence and Security Act of 2007.

[5]For explanations of recent Department of Energy innovation initiatives, see Kristine Johnson presentation in upcoming book, Charles W. Wessner, *Clustering for 21st Century Prosperity*, Washington, DC: The National Academies Press, forthcoming.

thinker." Bush thought the government should fund basic research but that applied technologies and product development should be left to the private sector. "His initial vision was that there should be a continuum, and that we in the United States should support that," she said. The approach was successful at the start and continues to serve as a model. "But the world has become more complex in the last 65 years," she said.

The process of innovation was revolutionized by the invention of the transistor, which led to the computer, Dr. Johnson said. Thanks to these breakthroughs, she said, "we can now model and simulate what we want to build before we build it." She noted that the Boeing 777 and some submarines have been designed and built without physical prototypes. "It's stunning if you think about that kind of change over the past 60 years," she remarked.

For a good model of how innovation works, Dr. Johnson recommended a book called Pasteur's Quadrant.[6] The book describes a model in which innovation is conducted along two axes. One axis represents fundamental research. The other represents research into applied technology. Louis Pasteur operated on both levels: He engaged in very basic scientific discovery as well as development of vaccines made possible by that research. At the DoE, the approach is to focus on "use-inspired" research, Dr. Johnson said. The agency attempts to be both at the cutting edge of fundamental discovery as well as at the frontier of applied research. "So we need new models for innovation," she said.

The DoE's Energy Frontier Research Centers fund innovation in research, Dr. Johnson explained. The ARPA-E program, by contrast, funds innovation in technology. "As an engineer, that is where I like to play," she said. The energy research hubs "fund innovation at scale."

Work in bio-fuels illustrates how these DoE programs support research at every level of the innovation continuum. The Energy Frontier Research Centers explore "ways to be biologically inspired by the way plants convert sunlight CO_2 water into energy and bypass having to grow, harvest, and gasify the plants to create fuel," Dr. Johnson said. ARPA-E explores ways to implement breakthroughs. The hubs "are trying to gather all of the best ideas together in order to reach commercial scale," she said.

---

[6]For explanation of the model of innovation based on the balance of basic and applied research, see Donald E. Stokes, *Pasteur's Quadrant: Basic Science and Technological Innovation*, Washington, DC: Brookings Institution Press, 1997.

This new approach to innovation is being "pioneered and led" by Energy Secretary Steven Chu, she said. "Part of our innovation is that we need to be open and collaborative, and together figure out the best ideas that will move our civilization and planet forward," she said. The approach used for bio-fuels is also used for "the entire search for the next generation of renewable energy."

Collaboration is vital if the DoE is to achieve its goal of doubling renewable electricity generating capacity and advanced energy manufacturing by 2012, Dr. Johnson said. The Recovery Act investments and incentives of $23 billion, combined with private capital of $40 billion, "will give us the fuel to help meet these goals," she said. Still, there is not enough funding from one source alone to make this clean-energy revolution happen.

To give one example, Dr. Johnson said she has been very close to the effort in hydro power. She noted that her father was a hydro power engineer and her grandfather was a hydro power engineer. "Now I am a hydro power engineer," she quipped.

The DoE spent $30 million, leveraged with about $100 million from private industry, to create 30 megawatts of new hydro power. Probably 30 gigawatts to 100 gigawatts of potential hydropower can be generated in the United States. But that would require $30 billion in investment, far beyond the agency's resources. "So we have to be clever about how we make those projects happen by using policy instruments," she said.

Dr. Johnson noted that the United States has 79,000 dams, but only 2,200 of them produce electricity. In some cases it may make sense to remove a dam and have a small hydropower in that river. "That requires us to think strategically again over how we deploy this opportunity," The DoE is working with the Treasury Department to provide $2.3 billion in tax credits to assist such innovative activity.

The United States also is promoting conservation. The government has committed $5 billion to weatherize low-income housing. "This has the potential to save 20 percent to 30 percent on the energy bills in the hardest-hit families in the economy climate," she said. Buildings account for 40 percent of energy-use in the United States. Thirty percent of that energy can be cut through fairly simple energy-efficiency moves, such as installing more insulation and replacing incandescent bulbs with compact fluorescent bulbs. There also are incentives for more efficient furnaces and appliances. The federal funds have been matched with $3 billion from state governments and another $3 billion distributed to 2,300 cities, counties, territories, and Indian tribes. Also, the DoE is creating a regional innovation hub for energy-efficient building technologies.

President Barack Obama expressed America's commitment to the environment is a speech in Prague on April 5, 2009, Dr. Johnson noted. The President said: "To protect our planet, now is the time to change the way that we use energy. Together, we must confront climate change by ending the world's dependence on fossil fuels, by tapping the power of new sources of energy like the wind and sun, and by calling upon all nations to do their part. And I pledge to you that, in this global effort, the United States is ready to lead."[7]

The United States and China have a long history of working together, Dr. Johnson noted. "We are very proud of that history," she said, noting that the two nations currently are working through the bilateral strategic dialogue on subjects such as bio-fuels, wind, and transportation.

One important example of such cooperation is a partnership involving the DoE's Argonne National Laboratory, the U.S. Environmental Protection Agency, the Chinese Academy of Sciences, and U.S. and Chinese universities to model regional and local air quality. The program provided information to Beijing officials to improve conditions for athletes and spectators at the 2010 Summer Olympic Games, she said.

Dr. Johnson noted that she attended the Beijing Olympics, and wished to compliment China for the performance of its female field hockey team. Dr. Johnson, a former field hockey player herself, attended several Olympic matches. "If you can imagine sitting in the stadium at dusk and seeing a crystal clear sky," she said. "It was a perfect day for hockey. It was really inspiring."

The most important new collaboration is the U.S.-China Clean Energy Research and Development Center. The center was announced in July. The goal is for both the United States and China to each invest $75 million over five years in three areas—energy-efficient buildings, vehicles, and carbon capture and sequestration for coal. "I was pleased to have a role in planning this center," she said. "And I'm looking forward to working with our colleagues in the United States and China to make this an exemplar of international cooperation in solving the global climate challenge." She thanked her Chinese counterparts for their help.

---

[7] See The Office of the White House Press Secretary, "Remarks by President Obama," Hradcany Square, Prague, Czech Republic, April 5, 2009.

Renewable Energy Policy in China

Ren Weimin
National Development and Reform Commission

The growing pressure of energy demand is an important topic at the Academy of Macroeconomic Research, at the NDRC, said Mr. Ren, the unit's deputy director. "Developing clean energy is an inevitable choice that China will make," Mr. Ren said. By 2020, non-fossil fuel is expected to account for 15 percent of China's primary energy consumption.

These pressures will continue to mount as China seeks to attain its economic development goals. China already is the world's largest developing nation, Mr. Ren noted. The nation's long-term economic blueprint calls for becoming a "comfortably well-off society" by 2020, "more or less realizing industrialization" by 2035. By 2050, he added, China should be a "medium-level developed nation," but not yet reaching the level of the United States.

Many problems must be solved for China to meet its future energy demand, Mr. Ren said. In 2005, he noted, China already had reached very high levels of energy production and demand, consuming 2.2 billion tons of coal equivalent (TCE) worth of primary energy in that year. In 2009, consumption reached 2.8 billion TCE. Demand is expected to hit 3.1 billion TCE in 2010, 4.5 billion in 2020, and about 6 billion TCE in 2050. These required production level will be "gigantic," he said.

China currently is a major producer of fossil-based fuels. In 2009, produced about 3.05 billion tons of coal, 189 million tons of crude oil, 85 billion cubic meters of natural gas, Mr. Ren noted. These are the main sources of electrical power generation in China. "As you can imagine, as we look forward, the pressure on China's energy need is huge," he said. "So renewable energy is an inevitable choice for China."

In terms of renewable energy production, China's view is similar to that of many other nations around the world, Mr. Ren said. The country is seeking a mix of wind, solar, biomass, hydro, ocean, and geothermal power," he said. "This is our inevitable choice in our strategy in developing renewable energy."

The nation has substantial potential resources. Mr. Ren estimated that China's "basic conditions are good" to produce 1.7 trillion TCE of solar energy and the potential for 1,000 gigawatts of wind power. China also has "technically available exploit capacity" of 540 gigawatts of hydro power and a large amount of biomass and geothermal energy.

China has "a solid foundation for developing renewable energy," Mr. Ren said. One reason is that "Chinese leaders are very clear about their goals." In 2009, President Hu Jintao vowed that China will do its best to develop renewable and nuclear energy, aiming to have non-fossil energies account for 15 percent of consumption of primary energy by 2020.[8] "This is China's national policy, and also our solemn promise to the whole world," he said. By that time, he pointed out, China's annual energy demand will have more than doubled, to 4.6 billion CTE.

Meeting that target will be a major challenge. While China's conditions for renewable energy are favorable, Mr. Ren said, it is relatively expensive. As an illustration, he cited the cost differences between power from coal and other sources. In Xinjiang Province, coal is so abundant that electricity costs only 0.23 yuan (3.4 U.S. cents) per kilowatt.

Wind power, by contrast, costs 0.5 yuan to 0.65 yuan (7 cents to 9 cents). Biomass power costs 0.40 yuan to one yuan. With solar power, the price jumps to 1.2 yuan to 1.5 yuan (19 cents to 22 cents). "So you can see from these prices that we face a lot of hurdles," Mr. Ren said. "In order to develop renewable energy, we have to gradually reduce cost a great deal." Industrial enterprises that use energy inefficiently will be weeded out. Then it will be possible for enterprises to have both high productivity and energy efficiency, he said.

In terms of the energy mix, the Chinese government wants hydro power to account for percent to 8-9 percent of primary energy by 2020. China has 400 gigawatts of potential hydro power. "So far, we have only developed 50 percent of that," Mr. Ren said, with 196 gigawatts produced. "We hope the figure reaches 70 percent." The government hopes solar power will account for 20 gigawatts of electricity, biomass will account for 30 gigawatts, and wind for at least 200 gigawatts. Nuclear plants are expected to produce 70 gigawatts.

China set out its long-term agenda in February 2005, when it released the Renewable Energy Law.[9] The law was enacted in January 2006. Two years later, the long- and medium-term goal strategies[10] were published,

---

[8]Speech by Chinese President Hu Jintao to United Nations General Assembly, September 22, 2009.
[9]The Renewable Energy Law of the People's Republic of China was approved by the Standing Committee in the 14th Session of the National People's Congress on February 25, 2005. See
<http://www.ccchina.gov.cn/en/NewsInfo.asp?NewsId=5371>.
[10]For details in English, see "Medium and Long-Term Development Plan for Renewable Energy in China," National Development and Reform Commission,

Mr. Ren noted. The country already is making some progress. In 2008, annual utilization of renewable energy totaled 250 million CTE, or 8.6 percent of annual primary energy needs. Even though renewable energy declined slightly in 2009 as a percentage of total consumption, to 8.3 percent, output rose to 260 million CTE. "The reason for the drop in percentage is that total production and consumption of energy increased very fast," he explained.

In wind power, China now ranks as No. 2 in the world in terms of installed capacity. It has 80 manufacturers of turbines. China experienced the world's biggest boost in wind capacity last year, rising 124 percent in 2009. With 10,129 new wind-turbine generators added last year, capacity 13,803 megawatts. China has a total of 21,851 wind turbine generators with a capacity of 2,5805 megawatts. "We can say that China's wind power is developing rapidly," Mr. Ren said. The quality of turbines and assembly capacity of spare parts also have increased. "The current trend is very good," he said.

China also is the world's leading producer of solar panels, with total production reaching 4 gigawatts in 2009. Ten Chinese photovoltaic manufacturers are among the 30 biggest in the world, Mr. Ren noted. The nation now has 300 megawatts of installed solar capacity. In addition, China's is the world's biggest consumer and producer of solar water heaters. In 2009, 145 million square meters of water were heated with solar power in China.

China is producing fuel from all kinds of biomass as well, Mr. Ren said. National annual production of biogas reached 14 billion cubic meters in 2009, providing fuel for 80 million rural people. Annual output of bio-ethanol reached 1.65 million tons, and installed capacity of biomass power-generation plants across the country reached 3.2 gigawatts. About 5 million tons of biodiesel were produced. Geothermal heating also increasing, by about 10 percent annually, he said. By the end of 2008, geothermally heated water reached around 600,000 people in China.

Substantial barriers still must be overcome for China to meet its renewable-energy goals, Mr. Ren said. The two biggest are the costs of renewable energy and weak market competition, he said. Another major hurdle is that the market is still immature. There is a lack of knowledge of how to improve the market performance of renewable energy. There

PRC, September 2007 (<*http://www.chinaenvironmentallaw.com/wp-content/uploads/2008/04/medium-and-long-term-development-plan-for-renewable-energy.pdf*>).

also is "lack of broad social recognition" of the importance of renewable energies, he said.

Mr. Ren presented a long list of other shortcomings. A "weak industrial system and supporting capacity" is an obstacle to wider deployment of renewable energy, Mr. Ren said. The "policy system" is imperfect. There is little in-depth assessment and no "clear mechanisms guided by objective," he said. "Market monitoring mechanisms," policy coordination, public disclosure, legal frameworks, and security policies are weak, he said.

Research and development also needs to be strengthened, Mr. Ren said. Currently, China is weak in technological innovation. More R&D funds and a "clear roadmap" are needed. Some Chinese-made renewable-energy equipment is "technologically backward" and "lacks competitiveness," Mr. Ren said. The key technology, equipment, and raw materials must be imported. As a result, China still is far from "high-efficiency, large-scale development and utilization of renewable energy," he said.

Nevertheless, there is plenty of reason to push ahead. "The world financial crisis has provided opportunities for leapfrogging forward in renewable energies," he said. "We should have strategic positioning and strategic goals." Advocates of clean energy "should neither be overly optimistic nor overly pessimistic on this issue," he said.

China is not backing off of its long-term goals. It aims to produce and consume the equivalent of 700 million tons of coal in renewable energy by 2020, about 10 percent of primary energy, Mr. Ren noted. By 2030, the target is to have the equivalent of 1 billion tons of coal, 20 percent of primary energy from non-fossil sources, and one-third by 2050, which would be equal to 2 billion tons of coal. "By 2020, we hope renewable energy will be an effective supplement," Mr. Ren said. "By 2030, it will become one of the mainstream energy sources. And by 2050, it will become the main energy."

In terms of strategic importance, wind power ranks first, Mr. Ren said, wind ranks at the top. After that comes solar power, and then bio-fuels. With wind power, he explained, China is focusing on developing land-based systems and connecting them to the power grid. The government also wants to advance industrialization of wind turbines. In terms of solar power, China is focusing on thermal utilization technology that is connected with architecture and the "balanced development of the photovoltaic industry," he said.

China's strategy in bio-fuels is to focus on biomass feed stocks that do not require arable land or potable water. [11] That means "no robbing people of their grains, no grabbing land from crops, no snatching water sources from farm land, and no taking feed from livestock," Mr. Ren said. "As you know, China is a country that has just accomplished its goal of feeding people," he said. "So we will pay attention to liquid fuel that will not consume grain or food."

To go forward, China is developing a "comprehensive policy and institutional framework" for renewable energy. This involves strengthening laws and establishing a "very rational structure" for government policy. "Economic and industrial policy should be compatible with energy policy," he said. Another element of the plan is to nurture talent. China needs to "foster people who have the ability to development renewable energy and to make decisions," Mr. Ren said.

By implementing these measures, "China can improve its level of development for renewable energy to reach the goal that we promised to the world," Mr. Ren said.

---

[11]The conversion of agricultural land to grow crops for bio-fuels has been blamed by some for food shortages and rising prices for crops such as corn, commonly used to produce ethanol.

PANEL II
INNOVATION CLUSTERS AND THE 21ST CENTURY UNIVERSITY

Moderator:
Carl Dahlman
Georgetown University

This symposium offers "a fantastic opportunity to have a lot of exchange and understand a lot of the challenges we face," said Carl Dahlman, a former World Bank economist who has done extensive work on China and India. He said he hopes "we also will be able to come up with some very concrete things for work in the future, including very specific collaborations."

To put the panel discussion on university innovation clusters in context, Dr. Dahlman noted that a university has three big roles. Its first mission is to train high-level human capital, "which is important not only for science and technology but also more generally for managing economies," he said. The second is advancing knowledge. The third is applying knowledge from universities, or in other words to transfer technology.

China has made very rapid progress in higher education, Dr. Dahlman noted. Enrollment rates have risen from 2 percent in 1980 to 23 percent today. Now, China has more people in universities, 25 million students, than the United States, with 17 million. Forty percent of Chinese college students study math, science, and engineering. "That is a very impressive accomplishment," he said. Dr. Dahlman also noted that many Chinese students study in foreign universities. The largest number of foreign students in the United States is from China, "so we have a lot of exchange that way," he said. "Now we have to see how we can get more students from the United States into Chinese universities to further our understanding."

China is the world's third-largest spender on research and development measured in terms of purchasing-power parity, Dr. Dahlman added, and may soon surpass Japan as No. 2 within five years. China also spends a greater portion of its R&D money in universities than most other nations, including the United States.

In terms of transferring knowledge, China has made "dramatic progress" in setting up all kinds of science and innovation parks, Dr. Dahlman said. "I think we are very lucky to have with us one of the key people behind that," referring to speaker Lou Jing of China's Ministry of Education. The panel also featured presidents from two universities. Charles Vest, president of the National Academy of Engineering, also is president emeritus of Massachusetts Institute of Technology. Dan Mott from the University of Maryland, that is very active in technology transfer. The panel also includes a presentation by Ginger Lew of the Obama Administration on setting up university innovation clusters in the United States. "We are going to have a very rich discussion," he said. He urged each of the speakers to think of concrete areas and specific projects for further collaboration.

Universities, Science Parks, and Clusters in China's Innovation Ecosystem

Lou Jing
Ministry of Education

Universities play a very important role in China's strategy to build an innovation society, explained Ms. Lou, deputy director of the Ministry of Education's Department of Science and Technology.

China has an "ecosystem of innovation that is diverse, stable, and self-adjustable, and flexible," Ms. Lou said. China's goal is to be a leading source of research and development to "promote the development of the economy and society of China and even help develop global science and technology and world civilization," she said. Universities are an essential element in this innovation environment. Other innovation communities in this environment include the Chinese Academy of Sciences, the Chinese Academy of Social Sciences, corporate R&D departments, research institutes specializing in economics and in social development, and the Chinese research organizations of multinationals such as IBM, Intel, and Cisco. This ecosystem also includes public-service organizations that evaluate patents and provide intermediary services. Ms. Lou described this environment as a "system of innovation with Chinese characteristics."

The foundation of an innovation environment is a "knowledge innovation system that organically combines scientific research and higher education," Ms. Lou said. "This system's core, breakthrough point is business-based, market-oriented, and comprised of industry, academia, and research." To produce distinct results, Ms. Lou said, the innovation system must "take into account different regions' respective characteristics and advantages." It also needs a "socialized, network technology intermediary service system," she said, which will "require additional effort" in China.

A "national innovation system with Chinese characteristics" must be comprehensive, she said. It should include scientific innovation, technological innovation, product innovation, industry innovation, system innovation, cultivation of innovative talent, and an innovative culture, Ms. Lou said. It also should be "networked, diverse, dynamic, open, and inclusive." Management and operation systems also are required.

Ms. Lou outlined the major tasks of a national innovation system. One is to strengthen original innovation, integrate innovation from different sources, and encourage "re-innovation" by improving on technologies introduced into the country. Another task is to "create a better environment to cultivate innovative talent and leadership, especially those with special insights," she said. "We also have to cultivate an innovation spirit and atmosphere in our entire society."

As elsewhere in the world, universities in China are assuming a greater role and mission. "We all know that in the 21st century innovation has become a driving force behind a country's economic development," Ms. Lou said. "Countries around the world are endeavoring to raise their ability to innovate scientifically and technologically, placing a high priority on cultivating talent and building energetic innovation."

The first mission of universities is "to serve as an engine or original source of a country's core competitiveness," Ms. Lou said. She noted that universities are involved in science and technology, education, the economy, and society. "Universities contribute greatly to the rise and development of a great power, and are closely connected with the country's industrialization and modernization processes," she observed.

Elite universities are the core of China's research establishment. They account for three-fourths of scientific theses. Of those university theses, around 75 percent come from the top 50 schools, she noted. A third role for universities is to "cultivate innovative talent," she said.

Universities have long been an essential force for innovation in China and "have solved or participated in solving major science and technology problems for China's economy. They also are involving in transferring and transforming technologies, Ms. Lou said. "Universities' continuous development in technology achievements brings a closer collaboration between academia, industry, and research institutes and demonstrates their potential for leadership," she said.

China's economic transition and modern technology trends makes contributions from universities even more important. Ms. Lou explained that China has growing demand innovative talent, new technology, and new knowledge. "The cycle of transforming knowledge into commodities has shortened," she said. "The relationship between science and technology innovation and national demand will be even closer, while demand will be higher."

Cultivating top research talent has become "our major task" in the past few years, Ms. Lou said. The goal is to "continuously provide innovation—not just fast but outstanding achievements," she said. Universities also provide technology transfer, strategic consultation, and other services. She noted that the 17th Party Congress called for establishing research bodies of high standards in universities to support a national innovation system that will increase China's competitiveness. "We want very concrete results," she said.

Universities are major repositories of high-level Chinese innovation talent, Ms. Lou observed. They employ 562 faculty that account for 40 percent of members of the Chinese Academy of Sciences and the Chinese Academy of Engineering. Universities also have 902 recipients of support from the National Science Fund for Distinguished Young Scholars[1] program, accounting for 63.3 percent of total. Universities host 73 "outstanding national innovation communities," 52 percent of the total.

The Chinese Ministry of Education operates a special "high-level innovation talent cultivation program," Ms. Lou said. The program funds 1,108 Cheung Kong Scholars,[2] 2,452 Outstanding Innovation Teams, 3,776 scholars classified as New Century Excellent Talents, and 126 innovation bases. "These figures show we are enhancing the cultivation of innovative talents and have seen some positive results," she said.

---

[1]The National Science Fund for Distinguished Young Scholars provides four-year grants of 2 million RMB to scholars, focusing on those under the age 45, who have made "outstanding achievements in basic research." Recipients select their own research direction. The goal is to "foster a group of prominent academic pacemakers in the forefront of world science and technology."
[2]The Cheung Kong Scholars program was established in 2005 by the Li Ka Shing Foundation and the Ministry of Education. It awards stipends on top of regular salaries and benefits to outstanding scholars in China, Hong Kong, and Macau.

The nation's research infrastructure is heavily concentrated in universities. Sixty percent of "national pilot laboratories" are on campuses, for example, as are 140 "national key laboratories," 63 percent of the national total, Ms Lou pointed out. Chinese universities also house 26 national engineering laboratories and 110 National Engineering Research Centers. There are 76 national university science parks with connections to more than 110 universities, she said. "We are promoting science parks to become important test beds for our talent to receive better training before they enter the work market," Ms. Lou explained.

Universities play a central role in national innovation tasks, Ms. Lou said. Universities are in charge of around 80 percent of research under the National Science Foundation's general programs, for example, including the major of "key" and "major" programs. Universities also run 40 percent of national high-technology research and development programs and 30 percent of research programs dedicated to "tackling key industrial problems of generic technology," she noted.

Universities also are becoming more important sites for applied technology. Ms. Lou noted that funding for converting research results into practical business applications is growing by 20 percent annually, and that approximately 40 percent of all university scientific research funding now come from business.

In terms of research results, universities generate more than 35 percent of patents, 60 percent of papers published in Chinese-language periodicals, and 80 percent of published papers in science and engineering journals, Ms. Lou noted. Universities receive more than half of all National Science and Technology Awards.

To develop China's innovation system, the government is putting a high emphasis on the continuity of scientific research, Ms. Lou said. It is taking into account the fact that technological innovation is developing exponentially and that emerging industries are becoming more concentrated. The government also is focusing on the interaction between science, technology, and policy.

Ms. Lou presented the following guiding principles of China's innovation strategy: To accurately position cultivation of talent and scientific research that serves society, to take into account the different advantages and characteristics of different universities so that they can collaborate, to boost original innovation by providing support for core technologies, and to combine, consolidate, and integrate activities to more efficiently allocate resources. The country also needs a more interdisciplinary approach to innovation, she said.

The government wants to better position Chinese universities, Ms. Lou said. One of the top objectives for the next five years is to establish a "schools-of-higher-education innovation system that fits in a socialist market economy and technological development patterns," she said. Another is to "markedly raise competitiveness and the quality of schools of higher education."

Universities and the U.S. Innovation System

Charles Vest
National Academy of Engineering

Dr. Vest's presentation focused on the fundamentals and key historical points of the development of American universities and the U.S. innovation system. Three years—1862, 1945, and 1980—were pivotal turning points, he said.

In the midst of the American Civil War, Dr. Vest explained, the federal government passed the Land Grant Act of 1862.[3] This legislation allotted a parcel of land to each state, the income from which was to be used to establish public universities to teach agriculture and "mechanic arts," or engineering. "This was the beginning of a great system of public universities that gave access to education and research-and-development work to a vast number of young U.S. men and women," he said.

The second major turning point came nearly a century later, in 1945. Until then, Dr. Vest noted, private industry funded almost all R&D at universities. The federal government was a "relatively small player," funding a modest amount of research in engineering, agriculture, and medical schools. "World War II changed everything," he said.

Science and engineering played a major role in the allied victory in the war. As peacetime approached, President Franklin D. Roosevelt wrote a letter to Vannevar Bush, a former engineering professor and entrepreneur who had played a major role in Washington mobilizing scientists, engineers, and industry for the war effort.[4] President Roosevelt

---

[3]The Morrill Act of 1862 (7 U.S.C. Sec. 301), also known as the Land-Grant College Act, gave each state 30,000 acres of federal land to establish colleges.
[4]Vannevar Bush (1880-1974) was director of the Office of Scientific Research and Development during World War II and is regarded as the architect of post-War U.S. science and technology policy. Dr. Bush maintained that the federal government should invest in basic scientific research, but that converting science into technology and commercial products was the role of private industry.

asked Mr. Bush how the scientific community could work in peacetime to secure the nation's economic vitality, health, and security.

Bush produced a famous report called *Science: The Endless Frontier*.[5] It made four primary recommendations that "seem very simple today, but were actually very radical in 1945," Dr. Vest said.

The first recommendation was that the federal government should view universities as the primary source of basic research in science, engineering, and medicine. Bush believed the government "should not start something new," Dr. Vest explained. The concept was that the government gets two things in return for each dollar spent on university research: The results of that research and financial support for "educating the next generation of scientists, engineers, and doctors," he said.

The Bush report suggested that federal research grants be awarded based on competitive merit. "It was to create what I would call a marketplace of ideas," Dr. Vest said. The report also recommended establishing the National Science Foundation, "which in its current form is one of the most important funders of basic research in U.S. universities," he said.

Under the system envisioned by Bush and his committee, federal funding of scientific research contributed to economic development through a "linear progression," Dr. Vest explained. Basic research led to applied research, which was followed by product development and then the introduction of goods and services into the market. The vision is dated, however. "Today's world is more complicated," he said.

Bush also believed in a *laissez faire* economic approach. "Support basic research, and the marketplace would decide which ideas are important and good. Private industry would move them toward products and services," he explained.

The third milestone in the development of the U.S. innovation system was the Bayh-Dole Act of 1980,[6] Dr. Vest said. This legislation allowed universities to own the intellectual property resulting from federally funded research in most cases. The U.S. government always gets free use

---

[5]See Vannevar Bush, *Science The Endless Frontier: A Report to the President*, Office of Scientific Research and Development, July 1945, Washington, DC: United States Government Printing Office, 1945.
[6]The Bayh Dole Act of 1980 (PL 96-517, Patent and Trademark Act Amendments of 1980), or the University and Small Business Patent Procedures Act, (PL 96-517, Patent and Trademark Act Amendments of 1980), gave universities control over their inventions stemming from federally-funded research.

of such intellectual property. By allowing universities to license and patent inventions, "this started a very different and increased relationship of universities to the private sector," Dr. Vest explained.

What does all this mean in practice? Dr. Vest stressed that the two primary missions of both public and private universities in the United States are education and research. "They have a third mission, which we broadly define as service to society," he said. "I think of all three of these mean that universities create opportunity—opportunity for graduates, opportunity for states and regions, opportunity for the world." However, "the role of service to society, particularly in economic development and technology transfer, definitely comes third," Dr. Vest stressed.

The so-called U.S. innovation system that evolved in this environment "frankly is not really a system," Dr. Vest said. "It is not designed or planned very explicitly." Nevertheless, the government, universities, and industry work together, he explained. "They create new knowledge and technology through research, educate young women and men, and create the next generation of knowledge and technologies," he said. "The marketplace then plays the role of moving these new ideas into the world as new products, processes and services."

Historically, this process has been a very decentralized, very loosely organized, and highly entrepreneurial system," Dr. Vest said. "Therefore, our innovation ecosystem has tended to vary from city to city and region to region, but always with these three components."

In balance, the U.S. system has been a great success, Dr. Vest said. Some economists estimate that more than half of America's economic growth in the past 60 years has been due to technological innovation, he noted, much of which came out of universities. Some of the most important innovations that have come largely out of universities include computing, the laser, the fundamentals of global positioning systems, numerically controlled machines, the organization and deployment of the World Wide Web, concepts of financial engineering, the genetic revolution, and much of modern medicine.

These were "big, earth-shaking changes and innovation, all of which had huge economic impact," he said. "But none were explicitly planned or envisioned in advance. So the role of fundamental research, freedom, flexibility, and entrepreneurship plays out in often-unexpected but very important ways."

Two other ingredients must be added to make the U.S. innovation system work, he said. The first is venture capital. "This risk-taking entrepreneurial approach to funding new ideas and new people to try to create new products, processes, and services has played an enormously important role, particularly in the last few decades," Dr. Vest said.

Clusters of innovation also have been very important. Dr. Vest explained that there essentially are two types of clusters—those that evolve naturally and those that are planned. The two most famous innovation clusters are Silicon Valley and Boston's Route 128. Neither was planned, he noted. "They came about because groups of bright people around universities and industry came together," he said. "The idea of venture capital developed, and great success ensued."

More recently, several planned and strategic innovation clusters have been created, Dr. Vest explained. Research Triangle Park in North Carolina is a good example. These clusters often began when large companies were attracted to a park to conduct research and development. These investments spawned smaller, more specialized companies in the area, "usually with one or more universities engaged," he said.

One feature of the U.S. innovation system is that every decade or so there seems to be a change in the way innovation works, especially from the perspective of large companies, Dr. Vest observed. In the 1970s, for example, innovation was dominated by central corporate research labs such as ATT's Bell Labs.

In the 1980s, as the U.S. lost competitiveness in manufacturing, "big companies reworked the way in which they did R&D and transformed it into a new kind of product development," Dr. Vest explained. In the 1990s, companies were performing better and were good at incremental improvements but realized they were not coming up with enough new ideas, he recalled. "So companies began purchasing their innovation by acquiring small start-up companies, frequently coming out of universities," he said. In the first decade of the new millennium, companies moved to the "open innovation" model, "which recognized, among other things, the global nature of innovation," he said.

What will be the new innovation system in the next decade? Dr. Vest said he doesn't have the answer, but had some observations. Life sciences and information systems "clearly will play a driving role in innovation in the next 20 or 30 years," he said. Another observation is that "we are going to be challenged in both of our countries to understand how to adapt our innovation system to large-scale challenges such as energy, climate change, food, and water."

The future of venture capital also is unclear, Dr. Vest said. "It is getting too risk-averse in the United States and is aggregating too much in a small number of large venture-capital firms." Another question is whether there will be a new disruptive technology to approach large issues such as energy, he said.

The most important question, Dr. Vest said, is "whether there will be a new enabling technology that will come along in the same way that information technology came along in the last century to change everything." A final question: "What does the globalization of research and development, of education, and a highly educated and innovative workforce mean for all of us?" Dr. Vest said he would leave all of these as questions for discussion.

Universities as Drivers of Growth in the United States

C. D. Mote, Jr.
University of Maryland, College Park

The University of Maryland at College Park illustrates the role American universities play in economic development, said Dr. Mote, the school's president and an engineering professor. He stressed, however, that education and research remain the university's primary missions.

To offer of a glimpse of the university's economic impact, Dr. Mote presented a few "factoids" that he said are "fairly typical" of U.S. universities:

- For every dollar the University of Maryland at College Park spends on faculty salaries, these faculty raise $3 in external research funding.

- Every dollar the state spends on the university generates $8 in economic activity.

- For every state dollar invested, the university raises $35 in development resources for small Maryland businesses.

- Every dollar in state investment has generated $200 worth of goods and services produced by university-supported companies over 25 years.

"You can see fairly quickly that the economic impact on the state by major research universities is very high," Dr. Mote said

It also is important to understand that American universities are "independent and free to engage in research and economic activity without permissions and without controls," Dr. Mote said. "This is true for private universities and for public universities too. For instance, a president of a U.S. university can to go the Ministry of Science and

Technology in China, meet with the minister, and arrange agreements that do not violate U.S. law without permission from the board of the university, or the governments of the state and the nation." This independence is a "fundamental contributor to the success of U.S. universities," he said.

A range of programs on the University of Maryland campus exemplify how "the spirit of entrepreneurship is embedded into the infrastructure of the university," Dr. Mote said. For instance, the Hinman CEOs program is a residence hall based program reserved for student entrepreneurs who want to start companies. In an average year, 17 companies are spawned in the dorm.

The Hillman Program campus works with Prince George's Community College to nurture entrepreneurs who tend to be older, in their 30s or even 40s, and have returned to college to help them to start businesses. "They come through the community college and transfer to the university to become entrepreneurs. They have ideas," he said.

The ASPIRE program, by contrast, helps engineering students who want to get jobs in private industry after graduating. ASPIRE, which is run by the A. James Clark School of Engineering and the Maryland Technology Enterprises Institute, gives students scholarships to work on real-world, faculty-supervised engineering projects with companies. The university's Smith School of Business, meanwhile, offers a business plan competition called Cupid's Cup. Students compete for resources to support starting their own companies by submitting business plans to veteran entrepreneurs serving as judges.

The university also participates in the Solar Decathlon, a competition administered by the U.S. Department of Energy. Twenty universities from around the world are invited to build solar houses on the National Mall in Washington, D. C. The University of Maryland team developed its "Leaf House" that operated off the electricity grid for eight days, including providing power for an electrical vehicle, "and had do a few other things to make it a little more of a challenge," Dr. Mote explained.

Companies launched through the University of Maryland include Alertus, a developer of emergency-warning systems. Another is Squarespace, a company that offers an environment for creating and managing Web sites and blogs.

The university also offers services to the surrounding community. For 25 years the engineering school has run the Maryland Technologies Enterprise Institute (MTECH) and the business school has operated the Dingman Center for Entrepreneurship. The two programs work together to create enterprises and offer services for new and existing companies.

The university also runs the oldest small-business incubator in the state and a "bioprocess scale-up facility." The latter unit "takes bench-top bioprocesses and turns them into commercial production-line processes," Dr. Mote explained. The state of Maryland funds both facilities.

The university works with industry as well. Maryland Industrial Partnerships is a program in which faculty are funded to work at companies to help commercialize their products. "Essentially, it is a state-subsidized consulting arrangement that has been extraordinarily successful and is being replicated around the United States," he said. Plus, there is a "venture accelerator," an organization that helps students and faculty speed up development of commercial products and companies. They receive training, introductions to financial backers, and mentoring.

The university also runs a "technology start-up boot camp." This is a weekend camp for people who want to start companies. Typically, these "boot camps" draw 500 to 600 participants from outside the university, Dr. Mote said. They teach "the good news and bad news of starting companies," including how to raise resources, why companies go bankrupt, and surviving the Valley of Death.

To help develop a pool of start-up capital, the university's business school has organized an "angel network." Angels provide early-stage funding, before good ideas achieve venture support. "It is a way for the business school to connect aspiring entrepreneurs with angels," he explained. And there is a business counseling program called Pitch Dingman, where the Dingman Center hosts a two-hour session where anyone can come, present ideas for new ventures, and receive feedback from experts who will then introduce them to angel investors if their ideas pass muster.

Dr. Mote estimated that all of these services and activities have generated $20 billion in economic activity over the past quarter century at a total cost to the state of approximately $88 million.

The university has launched or assisted a wide range of businesses. For instance, MTECH has helped develop commercial products with companies as diverse as toolmaker Black & Decker, the Quantum Sail Design Group, Hughes Network Systems, and engineering services firm Navmar.

Companies that have emerged from the university's incubator include two billion dollar companies: molecular diagnostic company Digene Corp.; Martek Biosciences Corp., a developer of nutritional products; plus the life sciences research firm NovaScreen Biosciences Corp. "They went through the incubator right at the beginning and used all of the services I described," Dr. Mote said.

The M Square Research Park, which is next to the university campus has received $5 million in state funding and $500 million in private investment, and will have about 2 million square feet of space and 6,000 jobs when built out.

In terms of the national domain, the University of Maryland interacts with federal laboratories, such as those owned by the Department of Energy. Its proximity to Washington, DC, also means the university is involved in a range of federal research initiatives and missions. Dr. Mote said the school gets around $500 million in federal research funding a year. Federal partners include the National Institutes of Health, the Smithsonian Institution, the National Aeronautics and Space Administration, and the National Security Agency. In some cases the university helps with government missions. Other times, the government supports university missions. "In other cases we work together on somebody else's mission," Dr. Mote said.

The National Oceanic and Atmospheric Administration is establishing a national center for global climate-change and weather-predication in the university research park. The DoE's Pacific Northwest National Laboratory, NASA, and the university's Earth System Science Interdisciplinary Center also are involved. Graduate and undergraduate students work in the center, which includes experts in geography, public policy, geology, and atmospheric and ocean sciences. Dr. Mote noted that the climate-change center also has worked with the Chinese Academy of Sciences and recently sent a team of 10 researchers to China.

Yet another federal partnership is with the National Institute of Science and Technology and is devoted to quantum physics. The agency is contributing some construction funds for a new physical sciences building on campus, and NIST scientists will work at the center. The campus also is home to an Energy Frontier Center and Physics Frontier Center. "These are ways in which the Department of Energy, National Science Foundation, and the university come together on research initiatives that are going to have great impact," Dr. Mote explained.

The university is also is engaged in a range of global activities. It has an "international incubator," for example, that has helped develop companies from Canada, Bangladesh, the United Kingdom, Russia, and other nations. There also is an "international research park," which is only available to international companies "to come and establish a foothold in the state of Maryland," Dr. Mote explained. The state contributes funding for the park.

The school has extensive relationships with China. There is the Institute for Global Chinese Affairs, for example, which trains Chinese executives. The first executives from China to visit the United States after the Cultural Revolution, a group from Suzhou, attended programs at the institute, Dr. Mote said. Some 3,000 Chinese executives have participated in the training programs, which last from two months to one year, he said.

The university's Executive Master's in Public Administration program, meanwhile, gives one-year degrees in public management to Chinese executives. The program was arranged with Secretary Liang Baohua of Jiangsu Province, and has trained 160 executives in five groups at Maryland's School of Public Policy. The executives finish their degrees after three months of additional study in China.

The university is home to a Confucius Institute, a Chinese program teaching and promulgating understanding of Chinese language and culture. The University of Maryland served as the pilot for the program in 2004, making it the oldest Confucius Institute in the world. There now are now 240 Confucius Institutes in 96 countries, including 74 in the United States. "It is a soft power, good feeling program, where each institute is free to set its own independent agenda and operation," he said.

Eleven Chinese companies have set up operations at the university's international incubator, Dr. Mote noted. Glodon Co., a developer of software for the construction industry, has been particularly successful. Formerly known as Beijing Grandsoft, the company went public, raising $2 billion. Within six months it was valued at $20 billion, he said.

Other Chinese companies in the incubator have included Wuxi TocaTek, DaSol Solar Energy Science and Technology, and Dimetek. Shandong Province set up a liaison office at the incubator. "The university has a role to play in facilitating this interchange," he said. "It shows what universities can do on an international scale to build enterprises."

In 2002, the Chinese government and Maryland set up a joint research park near the campus. When the park opened, Chinese Minister of Science and Technology Wan Gang travelled to the campus to attend the ribbon-cutting ceremony. Many Chinese companies with operations in the park were recruited through a series of meetings in Beijing, Shanghai, and Guangzhou, Dr. Mote said.

The University of Maryland has many other foreign partnerships, Dr. Mote noted. In Sierra Leone, for example, the university is involved in a health initiative. A Maryland graduate who now is a professional football player donated $2 million to set up a center to facilitate work.

To conclude, Dr. Mote noted that innovation has become a growing theme around the world. He cited Chinese President Hu Jintao, who in 2007 said that "the worldwide competition of overall national strength is actually a competition for talents, especially for innovative talents."[7]

The key to succeeding in innovation is leadership, Dr. Mote said. "Every innovative environment needs an innovation leader. Without a leader, and the ability to innovate within the infrastructure of an organization, it can never work. Leadership is everything."

U.S. Initiatives for Building Innovation Clusters

Ginger Lew
National Economic Council

The Obama Administration launched its regional innovation cluster initiative in the past year, explained Ms. Lew, a senior counselor to the White House and Small Business Administration on small-business issues. She acknowledged that the cluster concept itself is not new. "In Europe and Asia, regional innovation clusters developed with more of a top down process driven by government entities," she said. "The development of regional innovation clusters here in the United States has been much more on an ad-hoc, organic basis."[8]

Ms. Lew described cluster initiatives as consortia in which city and state governments and business, community, and educational leaders "come together to engage in smart economic growth strategies for a region." In the United States, she said, the process begins when a region assesses its local assets such as industry strengths, workforce skills, university research, and align those interests with future goals of key stakeholders.

One of the primary reasons for focusing on clusters as tools for economic planning and spurring innovation "is that businesses are no longer looking to locate in just one city," Ms. Lew said. "Rather, businesses are looking for the talent, infrastructure, and research capabilities that may be concentrated in a region that allows them to access what we call a more vibrant supply chain of vendors, services, and workforce." Another factor is that employees in the United States "no

---

[7]See October 2007 speech by Hu Jintao to the 17th CPC National Congress.
[8]For further explanation of U.S. innovation cluster policy, see presentation by Ginger Lew in upcoming book, Charles W. Wessner, *Clustering for 21st Century Prosperity*, Washington, DC: The National Academies Press, forthcoming.

longer work within defined boundaries," she added. "We are mobile. Sometimes we work virtually. And we certainly work across city lines and county lines."

A number of communities across the United States have undertaken efforts to develop innovation clusters, Ms. Lew explained. One major reason is that "economic studies have shown that clusters lead to higher paying jobs, more innovation, and more robust regional economies," she said.

Austin, Texas, which has focused on attracting a robust semiconductor industry, is the hub of one such regional cluster, Ms. Lew noted. Kansas has developed a strong regional aviation industry. Other communities have leveraged historical strengths to build new clusters. Ms. Lew cited Corning, New York, which parlayed its historical strength in glass into a fiber-optics industry. Seattle took advantage of its strong university system to develop a thriving bio-sciences cluster.

To illustrate the diversity of U.S. regional clusters, Ms. Lew displayed a map of the country by Harvard Business School's Institute for Strategy and Competitiveness.[9] The clusters on the map included oil and gas in Wichita, Kansas, entertainment in Los Angeles, and processed foods in Chicago. "When we talk about innovation, we oftentimes think about high-tech, such as nano-science, fiber optics, or whatever," she said. "But when you look at the map, you can see that you can have regional cluster activities in a broad range of industries."

The success of Kansas in commercial aviation is a good case study, Ms. Lew said. This industry employs 17.8 percent of all Kansas manufacturing employees and contributes 26 percent of manufacturing wages. What's more, the average annual wage of workers in the aviation cluster is $63,000—more than 50 percent above the average industrial wage in the United States. Between 2004 and 2014; the aviation industry is expected to create 4,450 net new jobs in the state.[10] "More importantly, it has increased the education level of the workforce," Ms. Lew said. "A number of the jobs now require a bachelor's degree and even a master's degree."

---

[9]The Institute for Strategy and Competitiveness, led by Michael Porter at Harvard Business School, has a project to map industrial clusters around the world. See
<*https://secure.hbs.edu/isc/login/login.do?http://data.isc.hbs.edu/isc/*>.
[10]See Center for Economic Development and Business Research, "Kansas Aviation Manufacturing," W. Frank Barton School of Business, Wichita State University, September 2008.

The Obama Administration is interested in innovation clusters because "we really see them as a way to encourage regional entities to collaborate to create new businesses and jobs," Ms. Lew said. "It also is a way to leverage federal programs." The United States has many federal bureaus and agencies that work in their own "silos," she explained. "But we are finding that activities across many agencies can be very complimentary."

One example of overlapping interests by federal agencies is clean energy. The Environmental Protection Agency has programs in clean water, for instance, that can be critical to efforts to develop alternative energy industries is certain regions. The EPA, DoE, and other agencies can work together. "This idea of leveraging federal dollars to be more impactful is a critical outcome we are seeking with regional innovation clusters," she said. "We are seeking what we call a multiplier effect."[11]

Unlike many regional cluster strategies in Europe, the U.S. model is very "bottoms up," Ms. Lew said. "We are looking to promote activity at the core regional level." Agencies in Washington, therefore, work with states that have comprehensive development plans. "We are looking for holistic, integrated solutions to building regional economies," she said.

Programs the Obama Administration has launched over the past year to promote regional clusters include:

- **Energy Regional Innovation Clusters (ERIC)**: Led by the DoE, federal agencies such as the Small Business Administration, the Department of Education, the National Science Foundation, and the Department of Labor are contributing funds and working with regional partners to develop innovation clusters in clean energy.[12]

- **Competitive Grants:** The U.S. Department of Agriculture, for instance, has launched a program to award planning grants to 12 regional bodies. The grants aim to "encourage rural communities to

---

[11]For elaboration on the philosophy of federal coordination on clusters, see Jonathan Sallet, Ed Paisley, Justing Masterman, "The Geography of Innovation," Center for American Progress, 2009. Also see Karen G. Mills, Elisabeth B. Reynolds, and Andrew Reamer, "Clusters and Competitiveness: A New Federal Role for Stimulating Regional Economies," Metropolitan Policy Program at Brookings, April 2008.

[12]The first Energy Regional Innovation Cluster is to focus on clean-energy technologies used in buildings. For details, see the Funding Opportunity Announcement for Fiscal Year 2010 on the DoE Web site. See <http://www.energy.gov/hubs/documents/ERIC_FOA.pdf>.

find new ways to draw on their core industries to attract more value-added business opportunities," Ms. Lew explained.

- **Small Business Loans:** The Small Business Administration will announce a competition to support ten regional initiatives across the United States to commercialize new technology.

- **i-6 Challenge Grants**. The Department of Commerce in May said it will award $12 million in grants administered by the Economic Development Agency to six teams across the United States with "the most innovative ideas to drive technology commercialization and entrepreneurship. The NIST also will contribute funds.[13]

- **2011 Federal Budget**: President Obama's budget for Fiscal Year 2011 includes more than $300 million in new funding for agencies such as the Department of Labor, the SBA, and the Economic Development Agency to assist regional innovation cluster initiatives.[14]

- **America COMPETES Act:** The most recent version of legislation to boost America's competitiveness in science and technology includes provisions for promoting regional innovation clusters.[15]

Ms. Lew outlined the structure of typical regional innovation clusters. At the core, she said, is the industry that a region or community identifies. Around that industry are suppliers, customers, and support industries. The rest of the ecosystem includes universities, community colleges, technical schools, federal agencies, labor groups, and non-government organizations, she explained.

In summary, the regional innovation cluster strategy of the Obama Administration has three core principles, Ms. Lew said. One is to

---

[13]See Announcement of Federal Funding Opportunity for i6 program at <*http://www.eda.gov/PDF/i6%20Challenge%20FFO%20FINAL%204-30-10.pdf*>.

[14]See Budget of the U.S. Government, Fiscal Year 2011, p. 20, <*http://www.whitehouse.gov/omb/budget/fy2011/assets/budget.pdf*>. For a brief analysis, see Mark Muro and Sarah Rahman, "Budget 2011: Industry Clusters as a Paradigm for Job Growth," Brookings Institution Metropolitan Policy Program, June 10, 2010, <*http://www.brookings.edu/opinions/2010/0202_fy11budget_cluster_muro_rah man.aspx*>.

[15]The America COMPETES Reauthorization Act of 2010 (H. R. 5116) passed the House of Representatives on May 28, 2010. It revises the original America COMPETES Act (P.L. 110-69). Despite being enacted on August 9, 2007, funding was never appropriated.

"encourage extensive collaboration at the regional level with business, university, and community leaders in public-private partnerships." The second core principle is to "encourage the collaboration and coordination of federal dollars." The third is to "cultivate an ecosystem to support the type of innovative, entrepreneurial clusters that will lead to new industries, new technologies, and new ways of doing things," she said.

Every county has its own approach to support innovation, Ms. Lew observed. The Chinese government supports the university systems. "The achievements (China) has achieved in a very short time are amazing," she said. The U.S. approach is more bottom-up with strong involvement from universities and some involvement at the federal level. "I think we can learn from both approaches," she said.

At the end of the day, however, "innovation resides in the mind of creative, smart individuals," Ms. Lew said. "They have to have the tools and skills. And they have to have the ecosystem to support that. But it only takes the curiosity of one person to come up with the next Baidu. That one person can launch a new industry."

As one small example of how individual curiosity drives innovation, Ms. Lew recalled her grandfather, who grew up in a small village in China. When he was young, her grandfather's job was to herd the family's ducks with a long bamboo pole. One of his favorite ducks sometimes strayed from the flock, and her grandfather had to find it. After straying several more times, the duck found a new source of food. So her grandfather allowed the duck to continue exploring to satisfy its curiosity. "Not only did it find additional new sources of food, but new sources of water as well," Ms. Lew recalled.

The responsibility of the federal government, university presidents, or the National Academy of Science "is to provide the type of support, skills and ecosystem that allows individuals to thrive."

Discussion

Moderator Carl Dahlman asked for further elaboration on several points made in presentations by the panelists. One is "the differences between the Chinese system, which is more top-down, and the American system, which is more bottom-up," he said. Another point is the "tremendous importance of having open systems." He asked Lou Jing and Charles Vest to comment on lessons that can be learned from each others' systems.

Dr. Dahlman also observed that toward the end of his presentation, Dr. Vest suggested the U.S. system may be a little dated given the new competition. He asked him to explain. Finally, Dr. Dahlman noted that "we are now in a system where we have much more global education, research and development, and flows of people." He asked how countries can adapt their innovation systems to this reality and how the United States and China can collaborate in new areas.

Ms. Lou disagreed that China's approach to innovation clusters is so top-down. "We have a number of nationally funded projects," she acknowledged. "These projects have played an important role in creating a platform at the national level."

But other models also exist, she said. Universities very actively support regional innovation clusters. "It is important to combine all participants in these regions," she said. "We also have a more horizontal development model. We have down to up as well as up to down."

Many places in China act as core centers to promote innovation in surrounding areas, she said. Examples are the Zhongguancun and Shandi districts in Beijing, the high-tech development zone in Shanghai, and the science and technology parks and research centers at universities and in provinces around the country.

Before responding to the question, Dr. Vest quipped that he had to apologize to Ginger Lew for his fondness of Peking duck. "I hope that by partaking, I don't stomp out some of the innovation in China," he said. Dr. Vest said he also agreed with Ms. Lew that innovation, clusters, job creation, and economic development "are not all about high tech."

Regarding cluster-development models, Dr. Vest said he is "a great believer in bottom up and open systems, by which I mean globally and regionally—not just nationally." He said he believes that "the most fundamental, true innovation is still going to come out of unexpected places and unexpected programs of basic research, not through planning."

Nevertheless, Dr. Vest predicted that this decade will be "one of re-balancing." The ideas of competition and cooperation on a global scale will be re-balanced, he predicted. "It also will be one of re-balancing the purpose of innovation and the nature of economic development."

Several fundamental issues will drive this change, Dr. Vest predicted. The "grand challenges" of the next century include energy, climate change, a global population that is approaching 9 billion people, and the rapid economic development of nations such as China. "These larger-scale issues that we simply have to resolve are going to affect the way that innovation works," he said.

An example of this shift in outlook is the strategy outlined by Energy Under Secretary Kristina Johnson for developing energy-innovation hubs. "This is a little more of a planned approach," Dr. Vest said. "It is a little more top-down, to define the problems we have to solve."

Universities' approach to basic research also will continue to change. "Universities should remain focused on discovery of new scientific knowledge, new technologies, and new processes," Dr. Vest said. "But I think they are going to be increasingly use-inspired." Work at the interface of life sciences and engineering for medical applications and new ways of producing materials is evidence of the new focus. "People are simultaneously exploring the unknown, but with a broad end-goal in mind," he said.

There are other signs of change in innovation systems, Dr. Vest said. For example, several "very interesting" new universities are being started around the world. One is Olin College near Boston. Others are Aalto University outside Helsinki and the new Singapore University of Technology and Design. "What they all have in common is an attempt to blend engineering and design in a very broad sense, running all the way from art and architecture to industrial design, with a good dose of economics," he said. "They all are searching for something new. I think they will be creating new kinds of people."

There also are new policy tools. In addition to the challenge grants Ginger Lew mentioned, there are U.S. inducement prizes such as those offered by the X Prize Foundation.[16] "Google the X Prize Foundation and you'll find some really interesting ways of driving innovation that are new and goal-oriented," Dr. Vest said.

In general, "we are moving into an era of what I would call brain integration," Dr. Vest said. "Somehow, in the coming decades, people all around the world, connected by huge computing and communication power, will start innovating collectively in ways we cannot predict. But I think this is why we have to maintain, in the near term, good people-to-people contact. It also is why I very much believe in openness of systems and science and engineering communication. Something new and exciting will come out of that, but I don't know just what it is."

Dr. Mote of the University of Maryland said he thinks leadership is one of the most important keys to innovation, even more important than the research topic itself. "In virtually every instance of successful innovation, you will find leadership in terms of inspiration, ideas, and in

---

[16]The X Prize Foundation is a non-profit educational organization whose mission is to "bring about radical breakthroughs for the benefit of humanity." It awards industry-sponsored prizes for innovators working on everything from genomics and automobiles to new spacecraft.

being able to mobilize focus on a topic," he said. "With the right leadership you can make marvelous things happen."

If one accepts that position, "you can structure innovation in the world in six layers," Dr. Mote said. The first layer is the individual. "An individual has to have innovative personality, focus, and capability," he said. The second layer is organizational—a company, university, or "just two or three people who come together and have an innovative idea." Sergei Brin and Larry Page, who started Google in 1998, are examples. "They didn't have a company, just two innovative individuals," he said.

The third level is regional—collections of organizations and individuals "that may not even have a specific focus," Dr. Mote said. The next is the state or provincial level. There needs to be leadership at the governmental level "that has the same level of authority and, possibly, inspiration." The fifth level is national. Leadership typically comes from the president or the presidential equivalents.

Finally, there is the global level. When one goes down the list of great international challenges—such as climate change, water, terrorism, and oceans—"all involve innovation on a global platform," Dr. Mote said. "That requires leadership on a global scale." The United States, however, is not well positioned to participate at this level. "We are very much bottom-up," he said. "We begin to run out of steam once we get to the regional level."

Efforts by the departments of Energy and Commerce to facilitate innovation clusters are "a marvelously good, important step," Dr. Mote said. This kind of federal collaboration hasn't occurred in the United States since before 1945, he noted. "At some point, the national piece will have to come into play so that we can take on these challenges. Otherwise, it cannot happen, because these things must be done through intergovernmental relationships."

Dr. Mote noted that nations with top-down innovation environments, such as China, Singapore, and Russia, "are all trying to work their way to the bottom, while the United States is trying to work its way to the top. I think the collaborations between us will help us get there, because I think the whole spectrum has to be covered for us to take on these big challenges."

PANEL III
ICT AND INNOVATION:
GROWTH ENGINE AND ENABLING TECHNOLOGIES

Moderator:
Dan Breznitz
Georgia Institute of Technology

Dr. Breznitz, author of the book *Innovation in the State*,[1] said this panel accomplishes "what we wanted to do all along, which was to have a representative from the Chinese side, the American side, and from a truly global company who will talk about the important issues of information and communication technology and broadband."

The first speaker, Chen Ying, is deputy director of the software department of China's Ministry of Industry and Information Technology. "He has been important in creating and implementing policies with regard to the software industry, and especially with a subject we have heard a lot about today—intellectual property rights," Dr. Breznitz said.

The next speaker, Eugene Huang, is at the White House Office of Science and Technology Policy and is senior advisor to the chief technology officer of the United States. "He has been deeply involved in the national broadband task force," Dr. Breznitz explained. Mr. Huang also served at the Federal Communication Commission and the Treasury Department and was secretary of technology for the state of Virginia.

While the first two presentations covered ways in which innovation helps economic growth, Dr. Breznitz said, the third presentation explained "how global companies manage to innovate in very, very different environments and very different countries." The speaker, Mark Dean, is vice-president of technical strategy and global operations for

---

[1] Dan Breznitz, *Innovation in the State: Political Choice and Strategies for Growth in Israel, Taiwan, and Ireland*, New Haven, CT: Yale University Press, 2007.

IBM Research. In this role, Dr. Dean is responsible for "setting the direction of IBM's overall research strategy across eight worldwide labs," he explained. Dr. Dean is an IBM fellow and has won many awards for innovation and technical leadership.

Impact of Broadband on Economic Growth and Productivity

Chen Ying
Ministry of Industry and Information Technology

Information and communications technology has played an increasingly vital role as a driver of global growth and in China's rapid economic development, Mr. Chen said. The industry has grown faster than others over the past 30 years. Its contribution to global gross-domestic product has risen by about a percentage point each decade, he noted.

The ICT industry will "be very important to growth of our national economy," Mr. Chen said. China's goals now are to improve the information-technology industry and maximize its potential. The government is setting priorities on different IT technologies, promoting a greater diversity of products, and "improving the infrastructure and applications of IT in different areas of our daily lives," he said. "It will create a great number of technologies and products, and will open up new markets. It will cultivate integration with new industries that have strong innovative abilities and with high value-added and new areas of growth."

Breakthroughs are occurring in areas such as the next generation of the Internet, new visual display devices, and digital audio equipment, Mr. Chen observed. Worldwide, governments are developing national strategies to target such opportunities, he added.

Over the next five years, information and communication technologies are expected to create no less than $5 trillion in global market demand and will lead economic growth. "Meanwhile, the manufacturing and service industries that are based on ICT industries are maintaining a growth rate of 30 percent," he said. These include electronic commerce, modern logistics, and outsourced software and services. ICT also is helping "optimize redistribution of resources worldwide." The growth of the industry also is creating more market opportunities for information-technology products, he said.

One top priority is to integrate information and communication technologies into Chinese industries. ICT can have a major impact in renovating and improving existing traditional industries, Mr. Chen explained. "Wider deployment of ICT can greatly help Chinese companies optimize the efficiency of human resources, their cash flow, and logistic flow and boost their productivity and economic efficiency," he said. "At the same time, it can decrease the consumption of natural resources, protect the environment, and realize sustainable growth."

A recent World Bank study highlighted the potential economic impact, Mr. Chen noted. According to the study, a 10 percentage point increase in broadband penetration rates can boost GDP by 1.38 percentage points in developing nations and by 1.12 percent points in advanced nations. One quarter of GDP growth in the European Union and 40 percent of productivity growth can be attributed to ICT, he said. The impact is felt in all industries in which management software programs such as ERP have been integrated into the entire manufacturing and operations process.

Better use of ICT can even streamline industries such as steel, chemicals, and furniture, Mr. Chen said, citing a recent European Union study. Information technologies are applied throughout material purchasing, R&D, design, transportation, sales, distribution, marketing, and customer service. "They have effectively increased productivity and decreased operating costs," he said.

The wide deployment of global electronic commerce saved an estimated $2 trillion in costs worldwide in 2009, Mr. Chen said. Handling orders and transactions by hand is eight to 18 times more expensive than orders processed through e-commerce. ICT has been especially important in improving the efficiency of corporations and industries during the recent financial crisis and global recession, he pointed out.

One major goal going forward is to optimize the integration of ICT into Chinese industries, Mr. Chen said. Several years ago, the government promulgated a national information technology industry strategy for 2006 through 2020, he noted.[2] Rejuvenating the ICT sector is a major goal. "These policy statements have helped us have a blueprint and optimize our national ICT industry," he said. It also has encouraged Chinese to "apply ICT in our daily lives so that we can improve the scale." Information and communication technologies also are seen as a way to improve the efficiency of China's overall economy.

---

[2]China's 11th Five-Year Plan (2006-2010) calls for 30 percent annual growth, reaching $125 billion in revenue in 2010, and for boosting exports by 28 percent annually.

In 2009, income from China's electronic information industry amounted to 10 percent of the nation's industrial production, Mr. Chen said. Despite the financial crisis, China's software industry is still growing at more than a 25 percent annual rate. In the words of former Chinese President Jiang Zemin, Mr. Chen noted, "the ICT industry has become a multiplier of economic growth, a transformer of development methods, and an accelerator of industrial upgrades."

Broadband Strategy in the United states

Eugene J. Huang
White House Office of Science and Technology Policy

One reason the Obama Administration has focused so heavily on broadband is that "we believe it is critical infrastructure to stimulate economic growth in the United States," Mr. Huang explained. "The important thing about infrastructure is that unless you use it, and use it in very interesting and innovative ways, it won't contribute to economic growth." The Administration, therefore, is focusing on the "entire ecosystem surrounding broadband and how we will use it in the future in the United States to stimulate economic growth," he said.

One of the first pieces of legislation signed into law after the Obama Administration took office was the American Recovery and Reinvestment Act, Mr. Huang noted. This legislation allocated $7.2 billion to the U.S. Department of Agriculture and the Department of Commerce to be used for grants to stimulate broadband deployment throughout the United States. The law also required the Federal Communication Commission to develop a national broadband plan and a "broadband map" of the United States.

The Department of Commerce has deployed $4.7 billion over the past year through its Broadband Technology Opportunities Program. The money was used to support broadband infrastructure, expand and enhance public computer centers, and encourage sustainable adoption of broadband service, Mr. Huang said. The other $2.5 billion has been used by the USDA to help deploy broadband in rural areas. "Much like China, infrastructure in our rural areas, especially when it comes to broadband, is lacking in comparison to cities," he said. "So there has been a concerted effort to make sure our rural areas have the ability to take advantage of broadband and the opportunities that broadband presents."

Mr. Huang explained that he had played a role in helping develop the national broadband plan while he was on the FCC staff. The plan focused not only on deployment of infrastructure, such as wireless systems or wired fiber-optic cables, or on how to get more people to subscribe. "It also focused on what we called 'national purposes,'" he said. "How do you use broadband? And how do you use broadband to promote economic growth and other key national priorities?"

The long-term goals of the National Broadband Plan[3] over the next 10 years are that:

- At least 100 million U.S. homes should have affordable access to actual download speeds of at least 100 Mbps and actual upload speeds of at least Mbps.

- The United States should lead the world in mobile innovation, with the fastest and most extensive wireless networks of any nation.

- Every American should have affordable access to robust broadband service, and the means and skills to subscribe if they so choose.

- Every American community should have affordable access to at least 1 gigabit-per-second broadband service to anchor institutions such as schools, hospitals, and government buildings.

- To ensure the safety of the American people, every first respondent should have access to a nationwide, wireless, interoperable broadband public safety network.

- To ensure that America leads in the clean-energy economy, every American should be able to use broadband to track and manage their real-time energy consumption.

What is clear about the goals, Mr. Huang said, "is that they are extraordinarily ambitious and, more importantly, are focused on how we use broadband for the future to promote economic growth."

Mr. Huang then displayed the first "broadband map" created as a result of the Recovery Act funding. The color-coded map depicts broadband penetration rates across the country, ranging from deep red for areas where zero to 10 percent have broadband access to deep blue for areas with concentration rates of 91 percent to 100 percent.

The deep blue areas primarily are on the East and West coasts and "in pockets throughout the United States that primarily are in urban areas," he said. "For the areas that are more red, you will see that broadband penetration generally is lacking in rural areas of the United States."

---

[3]See Federal Communications Commission, *Connecting America: The National Broadband Plan*, *<http://www.broadband.gov/download-plan/>*.

The $2.5 billion managed by the Department of Agriculture dedicated to rural areas is meant to address that deficiency. "We recognize that we need to continue to do a better job, focused along the lines of what the United States did in the 1930s when it determined it was a priority to get telecommunications distributed throughout the U.S.," Mr. Huang said. As a result of that commitment, he noted, the penetration rate of phone lines is close to 99 percent. "We hope to get there in the same way with broadband communications."

In terms of next steps, Mr. Huang noted that broadband was a key part of the Strategy for American Innovation[4] released by the Obama Administration in September 2009. In addition to calling for expanding broadband infrastructure, the strategy committed to assuring network neutrality to preserve free and open Internet access. To carry out these recommendations, the Administration created a broadband subcommittee of the National Science and Technology Council's Committee on Technology, Mr. Huang explained.

The Administration also has several programs aimed at using broadband infrastructure to promote economic growth and national priorities. For example, "there is a real focus to insure that smart grid deployment occurs throughout the United States," he said. Broadband is integral to that strategy. The National Recovery Act included $15.5 billion for smart-grid technologies and implementation.

The Recovery Act also included $19 billion in funds for health care information technology. Broadband "can be used to improve delivery of health care services and health care outcomes to the American people," Mr. Huang said. Yet another critical area involving broadband infrastructure is public safety communications, where he said the Administration will soon make some announcements.

The Administration also is interested in using broadband to more effectively deliver public services. They include areas like open government, expanding online service delivery, and integrating new media and social media. One of the Administration's key initiatives, he explained, is Data.gov,[5] a portal where the government can put data

---

[4]Executive Office of the President, "A Strategy for American Innovation: Driving Towards Sustainable Growth and Quality Jobs," National Economic Council, Office of Science and Technology Policy, September 2009, (<*http://www.whitehouse.gov/assets/documents/SEPT_20__Innovation_Whitepa per_FINAL.pdf*>).
[5]See <*http://www.data.gov*>.

online. There also is Apps.gov,[6] a portal for applications. "If you haven't seen some of these portals, I encourage you to visit them online," Mr. Huang said. "They are very, very innovative, and we hope they will pave the way for a new generation of public services online."

ICT Development in U.S. and Chinese Contexts

Mark E. Dean
IBM Research

It is becoming increasingly important that commercial enterprises "be allowed to work, collaborate, and innovate globally," said Dr. Dean, a 29-year IBM veteran. The United States and China both already benefit from globally integrated enterprises. And "most of the challenges and opportunities facing us can only be addressed with global collaboration and innovation," he said. "Innovation in isolation is not significant for a successful company or one country."

Globally integrated enterprises are a relatively new phenomenon on the world scene. By this, Dr. Dean said he does not mean companies that merely have operations and facilities around the world. "I'm talking about operating and innovating globally in a fashion that is inclusive and connected, rather than disconnected, across our borders," he said. "Many countries have yet to experience this. Thus, there is obviously some caution over what they wish to do."

IBM Research is a truly global organization, Dr. Dean explained. It has eight major labs—three in the United States, and others in Zurich, Haifa, Tokyo, Bangalore, and Beijing—that employ 3,000 researchers. About half of those researchers and 60 percent of IBM's 220,000 technical employees are based outside the United States. "So we have to work to not only innovate and integrate our operations globally, but also to maintain all of that in an effective operating environment," he said. "We cannot compete if we innovate in only one country because it is not effective."

Now IBM Research is looking to expand its presence further and explore new models of innovation. "Our old model of innovation has supported us well to date, but it is not sufficient," Dr. Dean said. "We are

---

[6]U.S. Chief Information Officer Vivek Kundra has described Apps.gov as a "one-stop source for cloud services." The portal contains business applications, cloud services, productivity software, and social media software. See <*http://www.apps.gov*>.

looking at how we can innovate even more broadly and in an even more integrated fashion."

IBM Research has a number of collaborative programs. For example, there are "co-laboratories," in which facilities around the world make five-year commitments to specific research projects. Other programs concentrate on deploying technologies around the world.

There also are collaborations with clients and joint-development projects in which IBM Research and other partners work together to accomplish something more effectively. Such partners include Kaiser Permanente, StatOil Hydro, the National Geographic Society, IDA Ireland, the Industrial Technology Research Institute in Taiwan, and Wanfujing in China. IBM Research also wants to establish a research presence in Africa and Latin America.

Each year, Dr. Dean is responsible for generating a global technology outlook for IBM Research. This full-year exercise explores the major technology trends that will affect IBM and its clients. Projections look out three to 10 years and are used to drive IBM's technology strategy.

One important factor in considering the future is that societal changes have a major impact. "We used to think that technology would drive society," Dr. Dean said. "That's not true. Actually, society chooses technology that it needs. We have embraced that."

One theme for 2010 is industrial transformations caused by the global economic downturn and shifting needs in energy and health care. Another broad area is "analytics and optimization," which includes responding to the mass digitization of the world. A third is transformation of software and services due to such rising phenomena as cloud computing. A fourth area is "systems and infrastructure," which looks at responses to the explosion in wireless traffic and needs to optimize costs, energy, performance, and time.

An example of how IBM Research applies some of the world's best minds to real-world problems is a large-scale simulation it is developing to analyze Tokyo traffic in real time, Dr. Dean said. IBM Research also is experimenting with using electric vehicles as storage facilities for the power grid in order to offset the lows and highs of electricity generation. In China, IBM has a deep analytical project to analyze supply-chain logistics in order to lower carbon dioxide emissions into the atmosphere.

IBM has a long history in China, Mr. Dean explained. The company had a presence in 1934. Like most multinationals, it left following the revolution and returned in the late 1970s when China opened its doors again. IBM's first sales office was an experiment. "We wanted to understand the culture, how businesses wanted to buy ICT technology, and how they wanted to work and innovate," he said.

In the 1990s, IBM began making strategic investments in China to take advantage of the tremendous opportunities. It began operating in several cities in multiple lines of business. In 1992, IBM began melding its China operations into its globally integrated enterprise model. "That has proven to be quite effective," Dr. Dean said.

IBM now has three R&D operations in China. They focus on three major areas—research, hardware development, and software development.

IBM's 200-engineer China Research Lab, established in 1995, collaborates with a number of Chinese partners. It primarily works on the company's Smarter Cities program[7] and software technologies. One of its "grand challenge" projects involves "the Internet of things."[8] IBM Research projects that for every person on the planet, there will be 1,000 things connected to the Internet. "If there are more than 1 billion people in China, that is a tremendous number of devices that you have to connect, manage, and provide information services to and from," he said. "How do you actually make that work?"

Other big IBM facilities are the China Development Lab, which has 5,000 engineers and was founded in 1999. It is one of IBM's biggest development labs, and focuses on software applications and services. The China Systems and Technology Laboratory, with facilities in Beijing, Shanghai, and Taipei, has 1,200 engineers specializing in systems. To be effective in China, it is important to do extensive field

---

[7]Smarter Cities is an IBM initiative that seeks to improve management of transportation, water, and other systems through next-generation information technology. See Suzanne Dirks and Mary Keeling, "A Vision of Smarter Cities: How Cities Can Lead the Way into a Prosperous and Sustainable Future," IBM Global Business Services executive report, IBM Institute for Business Value, 2009,
ftp://public.dhe.ibm.com/common/ssi/pm/xb/n/gbe03227usen/GBE03227USEN.PDF

[8]The Internet of Things refers to connected objects such as home appliances. The concept is attributed to the Auto-ID Center formerly based at the Massachusetts Institute of Technology. One good study on the topic is "ITU Internet Reports 2005: The Internet of Things," executive summary, International Telecommunications Union, November 2005,
<http://www.itu.int/osg/spu/publications/internetofthings/InternetofThings_summary.pdf>.

research "so that we can understand whether the solutions and technologies we are coming up with actually work," Dr. Dean said.

Open and global collaboration is key to IBM's research strategy in China. "There is not a single project we have across the research division that is isolated to a single country," Dr. Dean said. "We work to create a matrix that cuts across all our research labs." Most projects include personnel from multiple labs, including the China Research Lab.

The goal is to create technologies that have impact globally, not just in a single geographic location, he explained. Nor are research activities limited to information and communication technology. IBM researchers also work in unexplored areas of fields such as water desalination, advanced batteries, and solar cells, as well as software in businesses beyond where IBM presently is deployed. "Who knows what we might discover?" he said.

Partnerships have proved to be valuable to IBM, Dr. Dean said. Business partners account for 60 percent of IBM's revenue. IBM has 10,000 partners in 350 cities in China. They include CS&S, Futong, Digital China, Kingdee, and Yucheng Technologies. IBM offers 340 courses through its IBM Channel University. These courses reach 30,000 people through more than 3,000 business partners in 130 cities.

IBM also partners with Chinese universities. It has 100 joint labs and joint technology centers, Dr. Dean noted, and 80 special programs with 20 universities. So far, 860,000 students have been trained with IBM curricula, and 80,000 have been certified. IBM also has trained 6,500 teachers.

Dr. Dean agreed it is hard to understand how IBM operates globally and in an integrated fashion. "But we see this as key to our success," he said. "We work hard to avoid innovation in isolation, because that will create very narrow solutions that have very narrow upside potential."

To coordinate its global operations, IBM has executives responsible for the entire world. One chief technology officer runs global operations, for example. There also are global chiefs for human resources and legal affairs. "We don't replicate these activities in each country," he explained. "We have a common leadership."

A good illustration of IBM collaboration at work in China is the Smarter Cities project in Shenyang in the northeastern province of Liaoning. The mayor wants Shenyang to be China's first Smarter City, Mr. Dean said. IBM, the city government, and Northeastern University formed a five-year, $40 million partnership to achieve that goal.

The partnership is developing information and communication technology to manage systems such as water purity, energy, food safety, and integrated urban planning. "This is not only going to be an interesting and important exercise for IBM," Dr. Dean said. "We also think it will start an effort to replicate this approach in other cities, both in China and around the world."

Another major IBM effort in China is to develop technology to help utilities evaluate their networks and develop optimal investment plans to meet future demand, Dr. Dean noted. "The goal is services and technology that will allow us to look at each part of the distribution grid and the optimizations that can make that network operate more efficiently," he said. The target is to save 25 percent through such analytics. Potential partners are utilities in Tianjin, Chongqing, and Yunan Provinces.

Discussion

Dr. Wessner asked Dr. Dean to explain the major advantages of IBM's approach in China and the main challenges it faces. He also jokingly asked if IBM appeals to the "global government" if it has a problem in China, such as with intellectual property protection.

Dr. Dean explained that IBM has learned that Chinese businesses and individuals buy for different reasons. Therefore, IBM has "had to learn to build products and services that are a little more geared toward the challenges and opportunities that are there." The opportunities are with companies that want to grow 20 percent to 30 percent a year.

The big challenge for IBM "is to establish ourselves as an accepted entity in the region," Dr. Dean said. Because IBM has a U.S. headquarters, it is regarded as a U.S. company, he explained. But it also is global and integrated. "It is hard to express ourselves that way," he said. "We want to express ourselves globally, as a global entity. And we want governments to look at us that way."

But although IBM has made great progress in China and works closely with the government, "we're not necessarily viewed as a Chinese company, which can be a constraint in many ways. We would not like that to be an inhibitor," he said. "We would like to be viewed as an equal partner compared to indigenous companies, because we believe our investments will be on par with Chinese companies. I know Cisco experiences the same thing. We would like to be on a level playing field and bring innovation into China—not just innovate and pull out, but also bring. The total, I think, will be greater."

Fang Haiyang, director-general of the Shapingba District of Chongqing, raised a question about misuse of science. "I would like to ask how we can prevent the basic value of science from being misunderstood," he said. "The principle of science is to pursue the truth, explore natural rules, and enable people to make their lives more colorful and happier," he said. "In the process of industrialization and commercialization of every scientific research result, so troubles may also be brought to human beings." Mr. Fang noted that Alfred Nobel gained his reputation in science by inventing dynamite. "He brought much convenience, such as the use of explosives in infrastructure projects," he said. "But use of explosives in war brought disasters to human beings." Mr. Fang asked what the National Academy of Science thinks about preventing the abuse of science.

Dr. Wessner responded that it is a fair question. "One thing about the National Academy of Science is that it is full of scientists, which means it is full of opinions," he said. "It reminds me of an Israeli joke. If you have three scientists you have five opinions." Regarding negative uses of technology, Dr. Wessner said that "we look forward to (the Chinese) government's support in the United Nations with respect to some of the countries that seem to be inclined to create problems with technology."

He agreed that the two nations must work together to control technologies, and added that he was excited to hear about the potential to collaborate on transportation and energy technologies. In information technology, Dr. Wessner said "there is so much potential to rework the way we work, travel, and communicate." There also is potential in semiconductors. "As your government understands, perhaps better than ours, these lowly devices are vital to the information technology that has brought us such enormous progress in the last 30 years," he said. "I think companies like Cisco and IBM are showing a path forward on how to connect ourselves."

Dr. Wessner noted that the STEP Board convened a conference like this one with the Indian government two years ago. "We found that there were enormous interconnections we had not realized existed between the two countries," he said. Dr. Wessner noted that China is far ahead of India in terms of economic development.

Dr. Wessner said he was very impressed when a president of a Chinese university told him he had recently been vice-president of the University of Pennsylvania. "The sea turtle exchange is a source of value," he said. "I would hope over time that there is what the OECD calls 'circulation of high-value human capital.' In part, that is what this meeting is about, a chance to exchange ideas and to build relationships."

There is no simple answer to the question of misuse of science, Dr. Wessner said. "Since the beginning of time, science has been an opportunity for good and, alas, an opportunity for evil. We very much need your help in keeping this world in balance. We very much look forward to working with you to collaborate for a better planet."

Mr. Fang commented that he thought scientists should not be responsible for the industrialization and commercialization of their research results. Nor should they be responsible for investment and profit-making. "Government should be involved," he said. "I think the government should prevent technology from being misused and abused." He asked what role the government should play.

Dr. Dean said IBM has to deal with this question often. The National Academies brings transparency to science, enabling everybody to be part not only of discoveries but also their applications. "There always will be bad actors in the world," he said. "There is nothing you will do that can protect the world from a few people who will act badly. I am not sure we should constrain our exploration of science and its application by having people look over our shoulder."

While there should be checks and balances, Dr. Dean added that companies also are responsible for making sure they develop technologies in a way that that is safe for society. "IBM takes that very seriously," he said. "Policies governments put in place can help with that, but government alone can't carry the responsibility. No single government controls enough of the world to make sure science is not used in negative ways. We would constrain innovation if we have that in mind."

Moderator Dan Breznitz interjected with a comment on Dr. Wessner's joke. As somebody who was born in Israel, Dr. Breznitz said, he knows the ratio is not five opinions for every three scientists. "It is seven," he said.

Ren Weimin then asked about the Obama Administration's ability to fund its ambitious broadband plan. He noted that public funding in China is limited. "When will you be able to complete these projects?" he asked. "It's mentioned that the projects will need a lot of investment. In China, public funding is centralized. So I'd like to know how the project funding is structured here."

Mr. Huang of the Office of Science and Technology Policy explained that the information and communications technology plan is to be implemented over 10 years. It is hoped that by 2020, all of the long-term goals will be achieved.

Regarding funding, he acknowledged that the $7 billion provided in the 2009 economic stimulus act "is only a small drop in the bucket in comparison to what is necessary to meet all of the very ambitious goals."

According to some estimates, up to $300 billion would be required to achieve everything.

The bulk of that money will come from private industry, however. "The U.S. has a very robust system for competition in terms of broadband providers," Mr. Huang said. They include traditional telecom operators that are investing in fiber-optic capability and cable and wireless telecom providers that are investing to get broadband to the American public.

Mr. Huang noted that China has a unique opportunity to build state-of-the-art broadband: "There isn't the legacy of copper wires we have in our country, where we have telecom infrastructure going back to the 1930s," he noted.

When one studies broadband opportunities in the United States, one finds there is a clear opportunity for third- and fourth-generation wireless, Mr. Huang added. There also is opportunity for very high-capacity network connections for fiber-optic cable. "For us in the United States, it is not just a matter of picking one or the other technology," he said. "It also is to make sure there is an ecosystem that can leverage investment by the public sector, which is small, with investment from the private sector."

A member of the Chinese delegation asked how the United States initiated its smart grid program. She wanted to know if pilot projects were led by state governments, the federal government, or by a company, and whether pilots will be tested in one place first or launched nationally.

Mr. Huang explained that the federal role is limited. He said government investments in efforts such as smart grid and health-care information technology are a very small part of what is needed to fulfill deployment across the United States. The $15.5 billion mentioned in his presentation represents federal government investment in research, development, and very limited deployment, such as proof-of-concept demonstrations in communities. "We hope that once these particular initiatives demonstrate commercial value, they will be commercialized by the private sector," he said. "We hope that federal funding will help jump-start development and accelerate full deployment in the future."

Dr. Breznitz thanked the panelists, and said he hoped the discussion will stimulate dialogue for years to come.

PANEL IV
NEW FRONTIERS:
OPPORTUNITIES & CHALLENGES FOR COOPERATION

Moderator:
Bill Bonvillian
Massachusetts Institute of Technology

This panel addressed opportunities and challenges for cooperation between the United States and China in science and technology, said Mr. Bonvillian, who as director of MIT's Washington office manages the university's relationship with federal agencies and its role in national science policy. "As we all know, there is a remarkable amount of integration between our two countries on the commercial side," Mr. Bonvillian noted. "The two countries are major trading partners. And there are global enterprises that are remarkably integrated, though they still obviously have a distance to go."

This panel asked "whether we can cooperate more deeply than we are now on some very big societal challenges that we share," Mr. Bonvillian said. Issues discussed by the panel included health research, energy, water, information technology. Speakers also addressed the structure of the two nations' innovation systems themselves and how they can accommodate further collaboration.

Mr. Bonvillian said he hoped the discussion not only would explore what kind of cooperation exists now, "but more importantly what there could be." In the pre-commercial stage in particular, he asked, "What are the steps that might work? And could such cooperation be expanded and occur?"

The first speaker, Yang Xianwu, has worked on high-tech commercialization in the Chinese government since 1998, Mr. Bonvillian noted. Mr. Yang is deputy director of high technology and commercialization at the Ministry of Science and Technology. Over the years, he has worked with high-tech industry zones, business incubators, university science parks, and on boosting productivity. Mr. Yang has a particularly strong background in commercialization of information and space technology, he said.

Anna Barker has an extensive background in leading and managing scientific research, Mr. Bonvillian explained. Dr. Barker is deputy director of the National Cancer Institute at the National Institutes of Health. She also is deputy director for strategic initiatives in science. Dr. Barker "has done very interesting work on nanotechnology and its application to cancer," he said. She also has worked on the Cancer Genome Atlas Project. Dr. Barker also has been a scientist herself. Prior to joining the NIH, she led a large team performing cancer research at the Battelle Memorial Institute. Dr. Barker also has been CEO of a biotech drug-development company.

The third speaker, Robin Newmark, is director of the Strategic Energy Analysis Center at the National Renewable Energy Laboratory. "She has led very interesting work on the interrelationship between water and energy, including the impact on climate change, the de-nitrification of agriculture, and development of energy-efficient water-treatment technologies," Mr. Bonvillian said.

International Collaboration and Indigenous Innovation

Yang Xianwu
Ministry of Science and Technology

China has achieved remarkable economic progress since reform began and the nation opened up to the outside in 1978, Mr. Yang said. The economic has grown rapidly, living standards have improved, and China's overall strength has been greatly enhanced.

But China still faces problems, Mr. Yang noted. The level of Chinese industries remains relatively low on world standards. Most industries focus on manufacturing. During the financial crisis, many such enterprises closed. "The economic structure is unbalanced and improper," he said. "Also, our development is unsustainable." Manufacturing accounts for too high a proportion of the overall economy and the service sector too low. Energy consumption is relatively high, given the nation's GDP. "To tackle these problems, we feel we have to strengthen innovation, especially scientific and technological innovation, to transform China's economic development pattern and change the industrial structure," Mr. Yang said. "The ultimate goal is to make our nation an innovation country to match the level of countries such as the United States."

In 2003, China's central government began drafting a medium- to long-term science and technology plan from 2006 to 2020, Mr. Yang explained.[1] There are three major aspects to this plan. The leadership has identified 16 major national science and technology projects. The plan also identified priority areas for development, such as new energy sources, new materials, information technology, advanced manufacturing, biotech, and services enabled by information and communication technology.

The government is implementing a number of policies to achieve these goals, Mr. Yang said. It has been promoting science and technology through financial support, taxation policies, and support for research institutions, for example. Advancing China's innovation agenda isn't only the job of the Ministry of Science and Technology, he noted. The Finance Ministry also is active by funding basic research, such as through the National Natural Science Fund, which supports research based on scientists' own interests. The 973 Program supports projects with obvious application prospects, while the 863 Program backs frontier high-tech projects. Another program offers technology outreach for manufacturing and agriculture "to assume the good application of technology in areas that are directly related to peoples' lives," Mr. Yang said. Another program provides financial support to innovative small and midsized enterprises.

Many of these programs existed before the current medium- and long-term plan, Mr. Yang noted. "But since the plan, our financial investment from central and local governments has greatly increased," he said. Investments from the governments of Guangdong Province, Jiangsu Province, and Shanghai have grown several fold. Tax policies are another important tool, Mr. Yang said. The government offers generous breaks for high-tech enterprises and R&D investment.

While government investment in R&D has risen sharply, to 1.5 percent of GDP, Chinese companies have been slow to do so. Some 40 percent of R&D is funded by government, with the remaining 60 percent

---

[1]China's National Medium- and Long-Term Program for Science and Technology was issued on February 9, 2006, by the State Council. It calls for boosting research and development spending to 2.5 percent of gross domestic product and for science and technology to contribute at least 60 percent of the country's development. It also calls for reducing China's reliance on foreign technology to no more than 30 percent by 2020.

coming from companies and institutions. "I think this is low," Mr. Yang said. "Business and society should increase their funding for R&D."

Major weaknesses remain in China's innovation system, Mr. Yang said. The government is trying to stimulate more private research investment. For example, it is identifying promising high-tech enterprises and providing them with tax incentives. To qualify, these companies must be in high-priority areas and must devote a certain share of revenues to R&D, he said. Recipients also should have their own, proprietary patents and have developed good applications for their technology. For every yuan such companies invest, they will receive 1.5 yuan worth of tax incentives, he said.

In terms of policy, the government's goal is to "strengthen construction of basic conditions in priority areas," Mr. Yang explained. This includes establishment of national laboratories, high-tech research centers at universities, and research institutes with modern laboratory equipment. To accelerate the translation of research results into commercial products, China has been establishing more small-business incubators, university science parks, and high-tech industrial zones.

The government also is financing the establishment of laboratories, engineering centers and large science facilities. It is aiding projects that can serve as catalysts, Mr. Yang explained, such as university science parks, high-tech industrial parks, and innovation centers. "We're learning from the experience of Finland and America's Silicon Valley by establishing a large number of incubation centers to help scientists transform their research results and open their own small and medium-sized enterprises," he said.

The Ministry of Education has been promoting science parks since 1990. Most reputable research-oriented universities now have their own science parks, Mr. Yang said. There are 50 high-tech industrial parks throughout China, and more will be built. "They have become the most energetic areas in local economic growth and industrial development," he said. "We continue to promote cooperation and alliances between enterprises, research institutions, and universities so that they work together on R&D and transformation of the tech sector." Some technologically innovative enterprises receive no government money. "We just give them certain guidance and honor," Mr. Yang said.

Despite the growing emphasis on indigenous innovation, however, China still attaches great importance to international cooperation in science and technology, Mr. Yang said. "We have gradually realized that promoting international cooperation has played a very important role."

He noted that China has signed science and technology cooperation relationships with 152 nations and regions. It has sent science diplomats to 45 nations, and has signed inter-governmental agreements with 97 nations. China has joined 350 different international science and academic organizations, in which 265 Chinese scientists hold posts. China has participated in bilateral and multilateral programs, such as the Human Genome Project, the International Thermonuclear Experimental Reactor project,[2] and European Galileo Program.[3]

Few relationships have been more important than the one with the United States. "Cooperation with the U.S. has always been our priority," Mr. Yang said. The first Sino-U.S. Agreement on Science and Technology was signed Jan. 31, 1979, by Deng Xiaoping and President Jimmy Carter. This agreement has been extended every five years since, most recently in April 2006 by Chinese President Hu Jintao and U.S. President George W. Bush.

This partnership has achieved concrete results over the past decade. Mr. Yang noted that the United States and China have signed some 50 cooperation agreements over the past 30 years. They have covered fields such as agriculture, energy resources, the environment, and basic science, involving nearly all Chinese government agencies. Since 2004, the two nations have conducted scientific exchange programs each year.

The most recent agreement was the Protocol on Sino-U.S. Joint Research Center for Clean Energy, signed in November 2009. Both countries have promised they will provide the same amount of money, $150 million over the next five years, he noted. The protocol calls for collaboration on clean water, coal, automobiles, and energy. "This has historic significance," Mr. Wang said. "In the past, cooperation mainly focused on exchanges of personnel. This is the first time both governments donated directly to some joint development programs."

In 2009, Mr. Yang explained, China and the United States celebrated the 30th anniversary of scientific and technological cooperation. President Barack Obama has spoken highly of bilateral collaboration, he noted.

---

[2]The International Thermonuclear Experimental Reactor (ITER) project is a $12.8 billion multinational research and development program intended to develop fusion as an energy source.

[3]The European Galileo Program is building a global navigation satellite called Galileo that is meant to be an alternative as well as complementary to U.S. and Russian global positioning systems. The European Union and the European Space Agency are leading the multibillion-dollar project.

Despite these successes, significant problems still hold back collaboration in science and technology. "We have noticed some areas for improvement," Mr. Yang said. From the Chinese perspective, he noted, "we feel that in enhancing Sino-U.S. cooperation, both governments should deepen their commitment to investments." He also noted that the United States still places some restrictions on exports of high-tech products to China. In addition, high-level personnel from China continue to encounter unpleasant experiences in obtaining visas to the United States.

Mr. Yang acknowledged China's indigenous innovation "cannot be separated from international cooperation. "The U.S. is the most developed country in the world, while China is the most populated in the world and one of the countries with the fastest growth," he observed. "During the past 30 years, we realized that bilateral cooperation benefits us both. We're both pursuing a win-win result. So we can foresee that Sino-U.S. cooperation will develop further. For this, I am full of confidence."

Joint U.S.-China Medical Research Opportunities

Anna Barker
National Cancer Institute

Medical research is an area that is best positioned for collaboration between the United States and China, said Dr. Barker, the National Cancer Institute's Deputy Director. The National Institutes of Health already has a long history of working with Chinese researchers, especially in cancer. In fact, she said, some of the seminal research in cancer has been done as a result of the 30 years of collaboration between the two countries.

Why is collaboration important? One reason, Dr. Barker noted, is that the United States last year spent $2.5 trillion, equal to around 20 percent of gross domestic product, on health care. The health care reform bill passed in 2010[4] attempts to control future costs. "That is a big challenge," she said. Looking ahead, annual spending is expected to reach $4.5 trillion by 2018.

---

[4]The Patient Protection and Affordable Care Act (H.R. 3590) was signed into law on March 23, 2010. Among other things, it expands Medicaid eligibility, subsidizes insurance premiums, provides incentives for businesses to provide health care benefits, and supports medical research.

China is moving down a similar road as the United States, Dr. Barker said. Infectious disease is declining in China, while chronic diseases are rising significantly.

Another major trend in medical research is that there is an "important convergence" of advanced technologies, molecular biology and bioinformatics, which offers unprecedented opportunities for progress against diseases such as cancer, she noted. "When you think about great advances in science, it is in areas where the sciences come together," Dr. Barker said. "What we are seeing now in biomedical and medical research is that areas like computation, engineering, and physics, which we do not think enough about in biology, are converging in ways we haven't seen before. That represents significant opportunities for innovation."

Broadband offers another great opportunity to accelerate medical research as the infrastructure develops, Dr. Barker observed. "If you think about one of the areas where computation is really going to be called on to make a major contribution, it is in medical research," she said. The United States has made the sharing of digital medical records a major priority. "We'll see. It's very, very hard," she said. "These communities are very separated and, in many cases, siloed."

Cancer exemplifies the rise of chronic disease as a killer. "Cancer is a major U.S., and increasingly a global, health care crisis," Dr. Barker said. By 2020, cancer is projected to kill 10.3 million people globally. Such forecasts recently have been raised and are projected to reach as high as 20 to 30 million new cases per year on a global basis. It is an emerging crisis in China as well, "and will get much, much worse in the next 10 to 15 years, primarily due to the number of smokers," she said. There were 2.2 million new cancer cases in China in 2009 and 1.6 million deaths.

Dr. Barker displayed a chart showing how cardiac disease and cancer between 1973 and 2005 have emerged as the two top killers in China as the population ages. In 2009, cancer overtook cardiovascular disease as the biggest killer of people under the age of 85 in the United States. She predicted the same will happen in China.

The United States reports 565,000 cancer deaths each year. Another 1.4 million new cases are expected to be diagnosed in 2010. The United States spends $213 billion a year treating cancer. That will move much higher in years to come—approaching $1 trillion a year. New cancer cases are forecast to rise by 30 percent to 40 percent by 2020.

China is where the United States was around a decade ago. Cancer already is the No. 1 killer in Chinese cities and No. 2 in the countryside, Dr. Barker noted. It accounts for 25 percent of urban deaths and 21 percent of rural deaths. The aging population is a major reason these statistics will continue to rise. By 2035, 23 percent of Chinese will be at least 60 years old. As in the United States, obesity also is rising, with 23 percent of Chinese now considered overweight.[5] "That was not a problem in China until just very recently," she said.

Environmental and occupational hazards are other major cancer causes in China. There is a high incidence of liver cancer caused by Hepatitis B, a disease that has been conquered in the United States, Dr. Barker said. Lung cancer is perhaps the biggest danger in China. There are 350 million smokers, and many of them will get cancer. The pattern is similar to what the U.S. experienced, she observed. Lung cancer accounts for 29 percent of cancer-related deaths in the United States and has reached 22 percent in China.

While stomach cancer largely has been conquered in the United States, it remains a serious problem in China. The National Cancer Institute works actively with Chinese researchers on this problem. "We have learned a lot about stomach cancer from the Chinese, and we will learn a lot more," Dr. Barker said. She predicted stomach cancer will follow the pattern observed in the United States and be virtually eliminated in China based on treating patients for the associated infectious agent H. Pylori).

Serious bilateral research collaboration between the United States and China began in the 1970s. It was noticed that certain areas of China, such as near tin mines, had unusually high mortality rates. These areas represented hot spots in terms of environmental exposure, Dr. Barker explained. In 1979, the U.S. Department of Health and Human Services signed a memorandum of understanding with China. November 2009 marked the 30th anniversary of that agreement, and a symposium was conducted to discuss progress. A new memorandum of understanding is planned, she added.

Some of the National Cancer Institute's seminal epidemiology studies were done in China, she noted. "We learned a lot in China about cancer causation in some of these areas," she said. "It has benefited the world overall." One such study led to regulation of benzene in the United States, Dr. Barker noted. Important studies that advanced understanding of liver cancer and the impact of the environment on different cancers were done in China. The NCI' s Division of Cancer Epidemiology and Genetics teamed with China's Centers for Disease Control and Prevention, for example, to study the link between indoor cooking and lung cancer. The University of Washington and the Shanghai Textile

---

[5]Data from World Health Organization Health Information Profiles, (2008).

Industry Bureau studied cancer among textile workers. "Most of these studies have led to worldwide regulation of one form or another," she said.

China is currently making vital contributions in biomedical research, Dr. Barker said. In the past decade, the number of papers published by Chinese scientists in cancer research has more than quadrupled. China ranked second to the United States in published scientific papers in 2007, 2008, and 2009. The highest portions of these papers were in material science, chemistry, physics, and mathematics. The number of life sciences publications has risen exponentially, and many have been written with American colleagues, Dr. Barker noted.

The number of Sino-U.S. collaborations doubled between the periods of 1998 through 2003 and 2004 through 2008.[6] The National Cancer Institute works with the Chinese Academy of Medical Sciences, the Chinese Academy of Sciences, the China Center for Disease Control, and many major universities.

Cancer genomics is among the most exciting areas for investment in China. It now is possible to sequence cancer genomes, Dr. Barker explained. In the United States, the National Cancer Institute has launched a project called the Cancer Genome Atlas. The goal is to sequence genomes of all cancers. Chinese colleagues are collaborating on the project, she said.

Nanotechnology is another area were both the United States and China are making enormous investments, Dr. Barker noted. "We believe this is the area that is going to have one of the biggest impacts on the ability to detect and treat disease and deliver drugs," she said. "Nanotechnology will actually touch everything we do in medicine in the next 10 years."

In terms of cancer treatment, Dr. Barker said, China is a vital partner in clinical trials. "I should add that for cancer, nearly every major cancer center and hospital in China is being led by a scientist who trained at the National Cancer Institute and/or a U.S. medical school," she said. "We have a huge number of alumni in China. When we go to China, it is like going home—we are going to see people we know."

Cancer genomics will be important for developing new therapies and diagnostic technologies, Dr. Barker predicted. China's programs also are an opportunity to study rare cancers not found in the United States. The National Cancer Institute is working with China on brain, esophageal, gastric, and liver cancers, for instance. "We are looking at population

---

[6]Data from Global Research Report China (2009), Thomson Reuters.

differences as well as to build the whole area of genomics," she said. "China has led this area in the sequencing of the human genome, and I think will lead it with us for the next several years."

In addition to playing a big role in sequencing the human genome, China sequenced the rice genome. Its researchers were among the first to identify the SARS genome. In 2010, the Beijing Genomics Institute became the world's largest next-generation sequencing center, she noted. The National Cancer institute and BGI are researching brain tumors.

Research collaboration in nano-technologies also will grow, Dr. Barker said. China has 5,000 scientists at 50 universities in this field and 300 nano-technology enterprises.[7] It is second only to the United States in research publications. There are 20 Chinese Academy of Sciences institutes and 300 Chinese companies in nanotechnology. The Chinese government invested an estimated $240 million from 2004 through 2007, and local governments another $360 million.[8] "We are trying to build on our respective strengths in nanotechnology," she said. "This is a very strong collaboration." The third meeting between United States and Chinese medical researchers on this topic will be conducted in fall 2010.

The National Cancer Institute also wants to build on the collaboration with China in clinical trials and the study of environmental effects on cancers. "There is an opportunity to do this right from the beginning," she said. "We are working with the Chinese to build new clinical-trial systems with several hospitals in China."

The institute's goal is "to continue to expand our health care partnerships, and do that with a number of alumni in China and many of the post-docs that train with us here," Dr. Barker said. The institute has set up an office in Beijing, she noted, headed by Dr. Julie Schneider, who leads many ongoing collaborations.

In the future, medical and health care research "is going to be a very distributed enterprise," Dr. Barker predicted. "But I think it will be dominated by the U.S. and China because our countries are making the investments." Because of research in areas such as disease genomics, she predicted there also will be "a shift toward understanding the mechanistic causes of disease, which will lead us to a global understanding of how to prevent diseases like cancer."

Dr. Barker said she foresees such research leading to big changes in the health care system. "I think the knowledge base with bio-informatics and broadband will enable broad access where ever you are in the world

---

[7]Data from *Science* 309: 65-66, 2005.
[8]Ibid.

to the best available treatments," she predicted. Healthy populations will "define stability" in the future and will be critical to knowledge-based economies, she said. "China and the U.S. are in the best position to really dominate in medical research," she said, adding that she looks forward to the next 30 years of collaboration.

National Laboratories and International Cooperation

Robin L. Newmark
National Renewable Energy Laboratory

The National Renewable Energy Laboratories (NREL) is engaged in a wide array of clean-energy projects in China that reflect the laboratory's broad mission, Dr. Newmark explained. NREL, a U.S. Department of Energy national laboratory, is the nation's primary laboratory for renewable energy and energy efficiency research and development. In addition to conducting research and development on renewable energy and energy efficiency technologies, energy efficiency and technologies, NREL, based in Golden, Colorado, and tied to the Department of Energy, analyzes "the markets, financing, and policy mechanisms that will enable the great energy transformation we are all undertaking right now," she said. "Part of that work is the study of innovation."

Collaborations between government agencies of both nations range from research into broad national needs to narrowly focused partnerships with private companies to accelerate development of specific technologies. Many such projects stem from an umbrella agreement on clean energy negotiated through the U.S.-China Strategic Economic Dialogue.[9]

High-level engagements in areas of national interest include an Electricity Production and Transmission Action Plan, which explores best planning and management practices. A Clean and Efficient Transportation Action Plan exchanges best practices on new-vehicle technologies and design and management of transportation infrastructure.

NREL also aids two "eco-partnerships," which pair cities in the United States with cities in China for cooperation on specific clean energy and environmental objectives. Seven initial eco-partnerships were established under the U.S.-China Framework for EcoPartnerships, signed

---

[9]The China-U.S. Strategic Economic Dialogue is a framework agreement initiated by President George W. Bush and President Hu Jintao in 2006. High-level officials from both nations meet twice a year to discuss topics affecting economic relations between the nations.

in December 2008 under the S&ED Ten Year Framework for Cooperation on Energy and Environment. Six new partnerships have since been established. One is between Denver and Chongqing and focuses on transportation. A partnership between Greensburg, Kansas, and Mianzhu, the earth quake-stricken city in Sichuan Province, focuses on disaster recovery. Greensburg was devastated by a tornado and has since been rebuilt, Dr. Newmark explained.

The lab is part of several seven new Sino-U.S. clean-energy research cooperation centers initiatives announced during President Obama's mission to Beijing in the November of 2009. These initiatives span both conventional and non-conventional energy technologies. On behalf of the Department of Energy, NREL currently leads the U.S.-China Renewable Energy Partnership initiative for the United States, which interfaces with the Energy Research Institute under the National Development and Reform Commission in China. A wind-power project under the U.S. China Renewable Energy Program Partnership (USCREP) begins with national wind-energy deployment planning analysis and includes technical issues such as analysis of wake effects caused by downstream turbulence associated with turbine interference in large wind farms, and analysis of new wind resource assessment techniques based on SODAR technology. Another important component of the USCREP is cooperation for wind and solar standards, testing, and certification. Currently, U.S. and Chinese industry and leading experts are cooperating for new standards development in international forums such as the IEC, and are cooperating in comparing results between national testing centers and in general raising the level of international test center capabilities, and development of standards, Grid interconnection cooperation is also a high priority. Dr. Newmark explained. "Interconnection is very important for both countries as we continue to deploy renewable energy technologies and as we begin to incorporate that energy into the infrastructure of the grid," she said.

In addition to technical details, the wind partnership studies the economics of wide-scale deployment. As part of this project, NREL, LBL, and the Energy Foundation in Beijing Center for Resource Solutions of the United States is teaming with HydroChina, the State Grid Energy Research Institute, and the Meteorological Society to chart a cost supply curve for wind -power that depends on the potential installed capacity in different parts of China. These local cost-supply curves are used to map wind-power resources across the country that can be used to support development of a national plan. "Being able to link the geographic distribution of the resources with the investment potential for wind power deployment is where those resources are and how much money it will cost to implement wind power is really critical moving into the future," Dr. Newmark said. The same kind of methodology, she added, will be applied to solar power and its connections to China's existing transmission system. Other types of policy cooperation under the USCREP include: two workshops in 2011 to exchange information and experience on effective policy development to support renewable energy deployment, a workshop PV project evaluation and business model development to support market expansion, and training in several advanced analytical models at use at NREL for use in strategy energy analysis needed by the Chinese National Energy Administration.

A number of private-sector partnerships also are underway in clean energy. Chinese companies increasingly are investing in the U. S., Dr. Newmark noted. Duke Energy, for example, is discussing a U.S. wind-power plant with China's Huaneng Power International. Duke also is developing solar projects in the United States and joint technology in bio-fuels, clean coal, smart grid, and other areas with ENN Group. The U.S. Renewable Energy Group has proposed a large $1.5 billion wind project in Texas with two Chinese companies. And AMSC Windtec of the United States is working with Dongfang Turbine Corp. on an offshore wind project that would be designed by the American Superconductor Corp. Dongfang would own the intellectual property, she said.

Consistent with increasing interest in expanding U.S.-China renewable energy trade, Chinese companies also are beginning to building U.S. factories to make renewable-energy equipment and would like to increase clean energy investment in the United States at the manufacturing and project development levels. This interest includes plants to make solar photovoltaic panels and wind turbines by such companies as Suntech and Goldwind in China. "The bottom line is that China is making investments in U.S. projects and manufacturing capabilities," she said.

There are public-private partnerships between the two countries as well, Dr. Newmark explained. One example is the U.S.-China Bio-Fuels Cooperation Program. The project program was begun by initiated by the DoE, USDA and Chinese NDRC-China Memorandum of Understanding on Cooperation in the Development of Biofuels. Partners include five U.S. DoE national laboratories, agencies of the USDA, and federal labs controlled by the U.S. Energy and Agriculture departments and several research institutes and universities supported by the Chinese companies SinoPec, PetroChina, CNOOC, COFCO and COFCO ZTE.

The bio-fuels program has four major research areas, each with very different mixes of players, Dr. Newmark said. A project on the supply and logistics of feed stocks, for example, studies the economics and technical solutions for supplying non-food feed stocks for cellulosic ethanol conversion. Two U.S. national laboratories and the U.S. Department of Agriculture, ORNL and INL, are teaming with a range of Chinese partners.

Another set of bilateral projects focuses on processes for converting feed stocks into bio-fuels. One project involves NREL, Tsinghua University in Beijing, and PetroChina and other future partners focuses on biochemical conversion processes for cellulosic ethanol production. . It utilizes NREL work on characterization to study the breakdown of enzymes and other parts of the conversion process. Another project focusing on thermo-chemical conversion processes is a partnership between the Pacific Northwest National Laboratory in Richland, Washington, and the Dalian Institute of Chemical Physics supported by CNOOC. They primarily study pyrolyosis and gasification processes to convert biomass feedstock to mixed alcohols and other biofuels. A partnership between NREL and the Chinese Academy of Sciences in Qingdao seeks to develop biodiesel from algae and green diesel, which is derived from plant oils.

Once technology is developed by these collaborations, the private sector must use it to develop commercial products. NREL is involved in three partnerships with private companies that work with Chinese research institutes. "It is a ground-level interaction," Dr. Newmark said. "Through this cooperation, we go from a government-to-government to research institution to a commercialization train in order to develop new bio-fuels," she explained.

A partnership between NREL and ENN Group Co. focuses on commercial production of solar power and algae biodiesel, for example. The company is using 5.2-square-meter amorphous silicon module manufacturing equipment acquired from Applied Materials. NREL also works with the Institute of Electrical Engineering of the Chinese Academy of Sciences to cooperate on solar PV device testing. . The project uses NREL's testing know-how to cooperate for standards and certification processes for Chinese products, which are prevalent in the global PV marketplace, Dr. Newmark said.

On the purely commercial side, the United States and China have launched an energy-cooperation program that "is very unique," Dr. Newmark said. The program is a consortium of 30 to 40 large U.S. companies that are seriously engaged in China's energy sector. It is supported by the U.S. Foreign and Commercial Service, the trade-promotion arm of the Department of Commerce. "This consortium has developed a communications network that makes it very easy for them to interact with their Chinese counterparts," Ms Newmark said. "It helps them know who to speak to and where the connections should be made for various energy topics."

In sum, Dr. Newmark said, both the public and private sectors are heavily engaged in research collaboration between China and the United States. "We see rapid growth in interactions," she said. "There are enormous opportunities for mutual benefit with innovations from both the U.S. and China." Although many challenges in the relationship remain, "our observation is that we are making progress," she said.

Discussion

Dr. Wessner asked Dr. Newmark to elaborate on the barriers in China she mentioned in her presentation.

One barrier is concern over intellectual property, Dr. Newmark responded. Whether it is transferred or created through innovation, multinationals worry whether intellectual property is protected and whether its "commercial benefits will be maintained." In recent energy agreements with China, intellectual property has been addressed more specifically up front, she noted.

Mr. Bonvillian asked if intellectual property also is a concern in medical research and whether there has been any progress.

Intellectual property is a major issue in medical research as well, Dr. Barker said. "As collaborations increase, especially in biotechnology, international distribution becomes an issue," she noted. "I think we are seeing progress, but we have work to do."

Mr. Bonvillian asked both speakers whether U.S. visas also are a major problem in medical research.

Obtaining visas for Chinese counterparts was a significant barrier for the first year after the Sept. 11, 2001, terrorist attacks, Dr. Barker said. In 2009, "I think we were able to get visas for nearly everyone who wanted to visit with us," she added. "I think it is getting much better now than it was."

Dr. Newmark said NREL works with a large number of foreign visitors and that it is relatively easy to get visas for them. She added that NREL does hardly any classified work, though. For other national energy laboratories doing classified research, there are more barriers for American visitors as well as foreign nationals. "Overall we have seen an improvement in conditions since 9/11," she said. "I think we will continue to see improvement with the tremendous focus on energy research and the need for the perspective of multiple stakeholders and other nationals." Still, visas remain a problem, she said.

Mr. Bonvillian asked Mr. Yang of the Ministry of Science and Technology to elaborate on the visa problem.

While he is not very well informed on the current situation, Mr. Yang said, he had experienced problems with transfer of technology from the United States. "These issues are familiar to everyone," he said. In terms of travel visas, the ministry has heard complaints. China has experienced visa problems in collaborative life sciences research and in areas where there is "some sensitivity," he said. But this hasn't been an issue in energy research, he said.

Mr. Wang responded to some of the comments regarding intellectual property protection. He noted that systems to protect IPR have been in existence for a long time in other countries. Since China opened its economy, "it has achieved tremendous progress" in establishing laws. Many IPR courts also have been set up in China to address disputes, he added, and there have been successful cases of prosecution for intellectual property theft.

Intellectual property protection, however, "is not only the work of the government," he said. "Enterprises should provide evidence of IPR infringement. With evidence, a court will make a ruling." Companies should not count on the government to conduct investigations on its own.

Mr. Chen of the Ministry of Industry and Information Technology said he had heard of many IPR infringement cases, mainly through the work of the State Intellectual Property Office. In terms of software, the Chinese government has consistently expressed to the public the importance of using legal software. He noted that China now has a copyright law and is implementing regulations to protect copyrights. "Many, many manufacturers have complaints about enforcement of copyright protection," he conceded. China has set up a judicial system to interpret these laws. "I am sure the government can do more in terms of establishing this environment," he said.

The government took an extra measure in 2006, Mr. Chen said. The Copyright Bureau, in collaboration with the Ministry of Commerce, issued a ruling that all computers manufactured in China install registered operating systems. No piracy is allowed for computers leaving factories, he said. "If copyrights are infringed, companies should seek legal means to protect their rights. But the government did extra work," Mr. Chen said. "I believe this cannot be achieved by many countries."

These efforts are producing results, Mr. Chen said. Each year, the government collects measurement data on the 22 largest computer hardware manufacturers. Legal operating systems are now pre-installed in 90 percent of computers released by manufacturers, he said. "We have the hard figures to prove that, at least at the operating system level, the piracy issues have greatly improved," he said. "Many of our colleagues may not understand the measures taken by the Chinese government. In my view, many of the figures that people get were not obtained correctly."

He cited the example of a study on copyright infringement and software piracy in China that was based on only 150 product samples. "In a country as large as China, they made of conclusion on software piracy based on only 150 samples. I think we can ask whether these figures are accurate."

Dr. Barker asked if China is experiencing success with its business incubators and whether the government is doing anything to encourage technology transfer through incubators.

China's system is different from that of the United States, Mr. Yang noted. In the past, R&D mainly was conducted by universities and research institutes. "Chinese companies paid attention to production and didn't care about R&D," he said. Research institutes even designed products and then tried to convince manufacturers to produce them. Now, "we encourage companies not to wait for research institutes to give the R&D," Mr. Yang said.

The government is establishing platforms for commercializing R&D and to encourage innovators to set up business, Mr. Yang said. Science parks and incubators are examples. The government is supporting the effort with policies, such as preferential treatment. "The main driver is the market, however," he said. "We learned from advanced countries."

China's first incubator was established in Wuhan in 1987, he noted. Now there are incubators in most major cities. Many were built after science parks were established to encourage researchers to commercialize innovations.

Another Chinese delegate explained that many state-owned Chinese companies are weak in research and development. So China has borrowed strategies from developed nations. "We realized what is required is not just innovation in technology but also innovation in services," he said.

There are some preconditions for development of a high-tech industry, he noted. It requires venture capital, for example. The Chinese government is developing the venture capital market. Next, China needs intermediate agencies. "The government realizes that services and commercialization of results are very important. We will make all efforts to accelerate commercialization of results as China tries to form a more market-oriented economy."

Jim Hurd of the GreenScience Exchange observed that a private incubator in Beijing was started by Kai-Fu Lee, the former president of Google China. A second development, Mr. Hurd noted, is the evolution of foreign limited-partner venture capital firms in China. A new law is under discussion in the Pudong district of Shanghai to encourage the venture capital industry. He asked whether these developments also may be a source of cooperation between the United States and China.

Dr. Wessner asked how much money the Chinese government is investing to support small and midsized enterprises. He noted that the United States has the Small Business Innovation Research program, which provides almost $3 billion a year for early-stage funding. Dr. Wessner asked whether China has a similar program, and if so at what scale.

China's central government has been investing in small and midsized high-tech companies since 2000 through a fund announced by former Premier Zhu Rongji, Mr. Yang explained. In its first year, the fund totaled RMB1 billion. The fund is managed by the Ministry of Finance. The Ministry of Science and Technology advises how to utilize the funds. Mr. Yang said he believes the fund has been increased to RMB 4 billion. However, this fund only represents the federal government's investments, Mr. Yang noted. Local and provincial governments also provide funds for small and midsized enterprises.

Xu Jing, deputy division director of the Tariff Policy Department of the Ministry of Finance, asked how the U.S. government decides how to make such investments.

Mr. Bonvillian noted that defining the U.S. government role is very complicated because there are "so many different models in so many different sectors. So there is no simple answer."

LESSONS LEARNED AND NEXT STEPS

Moderator:
Michael Borrus
X/Seed Capital Management

Michael Borrus of X/Seed Capital asked the speakers in the previous panel how they set priorities in health and energy and how the Chinese government selects sectors to support.

In medical research, U.S. funding decisions are "still driven by the quality of science being done by individual investigators and teams of investigators," Dr. Barker replied. Institutions such as the Massachusetts Institute of Technology, and many others, encourage the translation of basic science into new commercial enterprises that are funded by the private sector, she added. "Our priorities in medical research are focused on research and development funded by organizations like the National Institutes of Health, which invests approximately $27 billion in medical research each year," she said. "We invest that across a whole range of medical research, from basic research to translational research to clinical research. In the U.S. technology moves out of the laboratory into translation." Technology and intellectual property are transferred to companies that are primarily funded through venture capital. "But we still set our priorities based on the quality and importance of the science," Dr. Barker said.

To summarize, Mr. Bonvillian said, life sciences research is driven by a basic research agency, and basic research drives what follows. He asked how this works in energy research.

Dr. Newmark of NREL said energy "is further along the continuum" that Dr. Barker outlined for medical research. "Energy research now is driven more by national priorities, both for the development of new options for traditional fossil and nuclear energy and for the development and commercialization of renewable energy and energy efficiency technologies," she said.

119

Discussions are underway over how to achieve the goals and benchmarks Energy Under Secretary Johnson explained, such as clean energy attaining a certain percentage of energy usage by a certain date, Dr. Newmark said. "The national priorities drive investment that span all the way from basic to applied research to commercial scale deployment," she said. "That said, the innovations depend on venture capital to get to commercial implementation." The U.S. government focuses on everything from very basic research in materials and separation science to very large-scale, public-private investments for deployment of some of these novel technologies."

Of course, the market ultimately will determine which energy technologies succeed, Dr. Newark noted. "In the energy sector, innovation requires that market finance and risk be addressed." But public financial support is required because new energy technologies require large-scale deployment to study and certify their performance, she said.

Mr. Bonvillian noted that the energy sector represents a different model of government involvement, one that mixes public and private roles "and is more bottom-up rather than top-down. And it is driven by a societal mission."

Cathy Swain, assistant vice-chancellor for commercial development at the University of Texas System, commented that there are a range of public investment activities in Texas. She noted that the state legislature set up a Cancer Prevention Research Fund several years ago. The $3 billion fund invests $300 million in cancer research each year. Texas also has the Emerging Technology Fund, which does not distinguish among industrial clusters, Ms. Swain explained. It focuses on early-stage investment for commercialization opportunities, before companies seek venture capital. The Fund also takes equity in start-ups.

Two other programs are part of the Emerging Technology Fund, Ms. Swain added. One is the Research Superiority Fund, which establishes centers of excellence at universities. The other is the Research Matching Fund. The state is looking to turn that fund into a source of micro-lending to support development of proofs of concept.

Texas' activities highlight another feature of the U.S. public funding model, Mr. Bonvillian noted. "There is a growing role by our states in sponsoring innovation," he said. "That has been nascent for a long time but is starting to grow."

He asked how Chinese institutions make decisions on what to fund.

The United States and China share many similarities, Mr. Yang said. "We also have a competitive method," he said. Recipients of funding go through a "very careful and dedicated evaluation process. No single individual or group can have a say." Various experts assess the merits of projects seeking funding.

Xu Bin, deputy division director of the high technology department of the National Development and Reform Commission, noted that the government provides very small funding, and wants more venture capital to come into to the country to make investments.

Lou Jing of the Ministry of Education explained that a number of national innovation policies are being developed. They include short-, medium-, and long-term goals. The government has identified priority industries that meet national priorities. There also are incubator activities in industries with great growth potential.

These strategies are developed by commissions composed of experts from different areas, Ms. Lou said. They include personnel from the Chinese Academy of Engineering, the Institute of Social Scientists, and the Chinese Academy of Science. Government institutions evaluate the process. "We will listen to advice from the local level, and then evaluate whether it is in line with the nation's long, mid- and short-term objectives to determine what kind of deployment timeline is needed," she said. "It is very comprehensive. Right now, we think our measures are very systematic and scientific."

Dr. Breznitz of Georgia Tech asked about China's approach to "investing in what society and the state see as critical infrastructure." He noted that the United States sees broadband as critical infrastructure for many reasons. "While we have let the private market do the heavy lifting, many rural communities have problems getting access," he said. "That is where government steps in, to allow every citizen in the U.S. to have access to this critical infrastructure."

Wang Xue, deputy head of the Chengdu High-Tech Zone, asked about America's health care reform. He said it seems that the reforms are built on a solid foundation, and that China may be able to benefit from the U.S. example.

She also asked Dr. Barker to share some of her insights on China's experience with cancer. He noted that the National Cancer Institute has a great deal of knowledge about different cases of diseases in different areas. Mr. Wang pointed out that in Sichuan Province lung cancer is the No. 1 problem. However, half of these cancer patients do not smoke. "I wonder if you have been following up and know what the situation is in Sichuan," he said.

Mr. Wang also asked her opinion on drug costs. If a truck manufacturer produces higher quantities of trucks, he observed, their costs go down. "How do you make drugs more affordable?"

"I feel like I am sitting before Congress," Dr. Barker quipped. "I have to answer the same questions there." Health care is a critical issue because its share of the U.S. economy is nearly 18 percent and rising, she noted. "I would like to say we have reformed our health care system, but we have just begun. The health care reforms address some of the issues but not all," she said. The reform will provide coverage for some, but not all, of the 40 million people who are uninsured, she explained.

Regarding China's situation, Dr. Barker noted that migration from rural areas to cities will create greater health-care issues. She observed that most advanced treatments for cancer are performed in cities. In rural areas, traditional Eastern medicine still is widely used. Urbanization, therefore, will drive up health-care costs. "Cancer is a particular problem because it is extraordinarily expensive to treat," she said. "It also is an area where an enormous amount of discovery is going on."

Drug discovery also is very costly. Pharmaceutical companies invest $1 billion on average for each new drug that actually gets to the market, and the process can take around 15 years. "That is something that has to change," Dr. Barker said.

The high cost of research in the West, however, means there is opportunity for substantial collaboration with China. "From what I see occurring in China, I think you will have an opportunity to change this with us or even before us," she said. "You have not built yet built infrastructure built that is driving costs. We are seeing a lot of your enterprises develop very quickly now in health care. I think some of your approaches have an opportunity to inform us here and result in very dynamic partnerships. I think we will learn a lot from each other."

The United States is struggling with the cost of drugs, Dr. Barker noted. She said she favors the concept of formulary, in which health care providers specify what they will pay for prescriptions in order to control costs. It may be difficult for the United States to adopt such an approach to public health care in "our free-market system," she said, although the formulary approach is starting to be used in health care for the elderly. The government must take into account the financial requirements of pharmaceutical makers, Dr. Barker suggested. "It is very difficult to incentivize drug companies without a very substantial profit motive because they have to invest a lot of money to develop a drug," she said. "So I think we are all in this together."

Regarding the question about cancer in China, Dr. Barker said she is familiar with the problem but "that nobody actually knows the answer." She noted that not everyone who smokes develops cancer, due to genetic differences. "However, we think there is something more going on there," she said. "Some of our scientists are working with your scientists so we can better understand this complex outcome."

Dr. Wessner offered some observations regarding venture capital. Often, too much credit in the United States is given to the role of venture capital in the innovation system. He held up a Blackberry and noted that the company that developed it never received venture capital. Lithium-ion battery maker A123, based in Watertown, Massachusetts, funded its research with funds from the National Science Foundation and the Department of Energy. It also received Small Business Innovation Research loans to develop commercial products. The company received venture capital after it already had developed its products. "It is claimed to be a venture capital success," he said.

Recently, Dr. Wessner added, A123 received $200 million from the U.S. government to build a production facility. "So we have a free market economy that is not always so free and not always so market," he said. "It also is not driven by the venture industry. There are many sectors where venture companies just don't go." A good example is medical research. If a company discovers an interesting molecule, "they ask you to come back after you have finished Phase I trials," he said. "If you tell them you need money for the Phase I trials, they tell you to come back when you have completed the Phase I trials." Dr. Wessner said he thinks there is an "obsession" with venture funding. "It is a much more complex mosaic than that."

Michael Borrus concluded the session by noting that he was struck at how much China and the United States have in common, "whether it be reducing dependence on fossil fuels or the fact we face rising cancer rates."

Based on the three decades of cooperation in health care between the two countries, Mr. Borrus suggested three principles on what makes collaboration successful. One is that "if cooperation is to work, it much be based on an equal exchange," Mr. Borrus said. "Each side must give as well as get." The second principle is that conflicts, misunderstandings, and sensitive issues are inevitable. "But we need goodwill on both sides to work through those problems in order to maintain progress toward building a cooperative relationship."

The third point is that "the only way to cooperate is to cooperate," Mr. Borrus said. To borrow a Chinese proverb, "We need to cross the river between our countries by feeling for he stones," he said. "We need to try some things together, demonstrate mutual gain, and then turn those smaller-scale collaborations into larger collaborations."

III
APPENDIXES

AGENDA

18 May 2010
Symposium on Building the 21st Century:
U.S.-China Cooperation on Science, Technology, and Innovation
Lecture Room
National Academy of Sciences
2100 C Street, NW
Washington, DC

9:00 AM **Welcome**

Charles Wessner, National Academy Scholar and Director of Technology, Innovation, and Entrepreneurship, The National Academies

9:10 AM **Opening Remarks**

Alan Wm. Wolff, Dewey & LeBoeuf LLP, and Chair, National Academies Study of Comparative National Innovation Policies

Ren Weimin, Vice President, Academy of Macroeconomic Research, National Development and Reform Commission

9:30 AM **Building Global Partnerships: Opportunities in U.S.-China Cooperation**

Anna Borg, Principal Deputy Assistant Secretary, Bureau of Economic, Energy, and Business Affairs, U.S. Department of State

9:45 AM **Panel I: Building the New Energy Economy**

Moderator: Michael Borrus, Founding General Partner, X/Seed Capital Management

New Renewable Energy Initiatives in the United States

Kristina Johnson, Under Secretary for Energy, U.S. Department of Energy

Renewable Energy Policy in China

Ren Weimin, Vice President, Academy of Macroeconomic Research, National Development and Reform Commission

10:30 AM **Coffee Break**

10:45 AM **Panel II: Innovation Clusters and the 21st Century University**

Moderator: Carl Dahlman, Luce Professor of International Relations and Information Technology, Georgetown University

Universities, Science Parks, and Clusters in China's Innovation Ecosystem

Lou Jing, Deputy Director General, Science and Technology Department, Ministry of Education

Universities and the U.S. Innovation System

Charles Vest, President, National Academy of Engineering

Universities as Drivers of Growth in the United States

C. D. "Dan" Mote, Jr., President, University of Maryland, College Park

U.S. Initiatives for Building Innovation Clusters

Ginger Lew, Senior Advisor, White House National Economic Council

12:30 PM **Working Lunch in the Refectory**

1:30 PM   **Panel III: ICT and Innovation: Growth Engine and Enabling Technologies**

Moderator: Dan Breznitz, Professor and Director for Globalization, Innovation, and Development, Georgia Institute of Technology

Impact of Broadband on Economic Growth and Productivity

Chen Ying, Deputy Director General, Software Department, Ministry of Industry and Information Technology

Broadband Strategy in the United States

Eugene Huang, Senior Advisor to the Chief Technology Officer, White House Office of Science and Technology Policy

ICT Development in U.S. and Chinese Contexts

Mark Dean, Vice President for Technical Strategy and Global Operations, IBM Research

2:45 PM   **Coffee Break**

3:00 PM   **Panel IV: New Frontiers: Opportunities & Challenges for Cooperation**

Moderator: Bill Bonvillian, Director, Washington DC Office, Massachusetts Institute of Technology

International Collaboration and Indigenous Innovation

Yang Xianwu, Deputy Director General, High and New Technology Department, Ministry of Science and Technology

Joint U.S.-China Medical Research Opportunities

Anna Barker, Deputy Director, National Cancer Institute

National Laboratories and International Cooperation

Robin Newmark, Director, Strategic Energy Analysis Center, National Renewable Energy Laboratory

4:15 PM   **Lessons Learned and Next Steps**
          Moderator: Michael Borrus, Founding General Partner,
          X/Seed Capital Management

5:00 PM   **Adjourn**

B

BIOGRAPHIES OF SPEAKERS[1]

ANNA BARKER

Dr. Barker serves as the deputy director of the National Cancer Institute (NCI) and as the deputy director for Strategic Scientific Initiatives. In this role she has developed and implemented multi/trans-disciplinary programs in strategic areas of cancer research and advanced technologies including: the Nanotechnology Alliance for Cancer; The Cancer Genome Atlas (TCGA); and the Clinical Proteomics Technologies Initiative for Cancer. She participates actively in these programs and serves in a team leadership role for TCGA. Recently she led the development of a new initiative to develop a network of trans-disciplinary centers focused on the elucidation of the "physics" of cancer at all scales through the establishment of Physical Sciences-Oncology Centers. Dr. Barker has also led and collaborated on NCI's effort to develop contemporary resources for cancer research in the areas of biospecimens and bioinformatics (The Cancer Bioinformatics Grid) to support molecularly based personalized medicine. She serves as the co-chair of the NCI-FDA Interagency Task Force; the co-chair of the Cancer Steering Committee of the FNIH Biomarker Consortium; and oversees the NCI's pilot international cancer research programs in Latin America and China.

Dr. Barker has a long history in research and the leadership and management of research and development in the academic, non-profit and private sectors. She served as senior scientist and subsequently a senior executive at Battelle Memorial Institute for 18 years where she developed and led a large group of scientists working in drug discovery and development, pharmacology, and biotechnology, with a major focus in oncology and NCI-supported programs. She co-founded and served as the CEO of a public biotechnology drug development company and founded a private cancer technology focused company. She has served in numerous volunteer capacities for cancer research and advocacy organizations including the AACR where she led the Legislative Affairs Committee for ten years and was a member of the Board of Directors.

---

[1] As of May 2010. Appendix includes bios distributed at the symposium.

She has received a number of awards for her contributions to cancer research, cancer patients, professional and advocacy organizations and the ongoing national effort to prevent and cure cancer. Her research interests include small molecule experimental therapeutics, tumor immunology, and free-radical biochemistry in cancer etiology and treatment. Dr. Barker completed her M.A. and Ph.D. at the Ohio State University, where she trained in chemistry, immunology, and microbiology.

BILL BONVILLIAN

William B. Bonvillian, since January 2006, has been director of the Massachusetts Institute of Technology's Washington, DC Office. At MIT, he works to support MIT's strong and historic relations with federal R&D agencies and its role on national science policy. Prior to that position, he served for 17 years as a senior policy advisor in the U.S. Senate. His legislative efforts included science and technology policies and innovation issues. He worked extensively on legislation creating the Department of Homeland Security, on Intelligence Reform, on defense and life science R&D, and on national competitiveness and innovation legislation. He has lectured and given speeches before numerous organizations on science, technology and innovation questions, is on the adjunct faculty at Georgetown, and has taught in this area at Georgetown, MIT and George Washington. He serves on the Board on Science Education of the National Academies, and has served on the Academies' Committees on "Learning Science: Computer Games, Simulations and Education," on "Modernizing the Infrastructure of the NSF's Federal Funds (R&D) Survey," and on "Exploring the Intersection of Science Education and the Development off 21st Century Skills." He was the recipient of the IEEE Distinguished Public Service Award in 2007.

His book, with Distinguished Prof. Charles Weiss of Georgetown, entitled Structuring an Energy Technology Revolution, was published by MIT Press in April 2009, and is summarized on the MIT Press Web site. His chapter, "The Connected Science Model for Innovation," appeared in the National Research Council book, 21st Century Innovation Systems for the United States and Japan: Lessons from a Decade of Change (May 2009). His recent articles include "Stimulating a Revolution in Sustainable Energy Technology" (with C. Weiss) in Environment (July/August 2009); "The Innovation State" (July/August 2009), and "Power Play—The DARPA Model and U.S. Energy Policy" (November/December 2006) both in American Interest with the latter reprinted in the book Blindside (Brookings Press, Francis Fukuyama, ed.,

2007);"The Politics of Jobs" (2007), "Meeting the New Challenge to U.S. Economic Competitiveness" (2004) and "Organizing Science and Technology for Homeland Security" (with K.V. Sharp, 2002), all published in Issues in Science and Technology; "Will the Search for New Energy Technologies Require a New R&D Mission Agency?" (2007) in Bridges; and "Science at a Crossroads" (2002), published in Technology in Society and reprinted in the FASEB Journal.

Prior to his work on the Senate, he was a partner at a large national law firm. Early in his career, he served as the Deputy Assistant Secretary and director of Congressional Affairs at the U.S. Department of Transportation, working on major transportation deregulation legislation. He received a B.A. from Columbia University with honors, an M.A.R. from Yale Divinity School in religion; and a J.D. from Columbia Law School, where he also served on the Board of Editors of the Columbia Law Review. Following law school, he served as a law clerk to a federal judge in New York. He is a member of the Connecticut Bar, the District of Columbia Bar, and the U.S. Supreme Court Bar.

ANNA BORG

Anna Borg, a Minister-Counselor in the Senior Foreign Service, is Principal Deputy Assistant Secretary in the Bureau of Economics, Energy and Business Affairs as of October 19, 2009. She previously served as DCM at USOECD (2008-2009), DCM at Embassy Rome from 2005-2008, and as Chief of Staff to the Under Secretary for Economic, Business, and Agricultural Affairs at the State Department from 2004-2005. She also served as Deputy Assistant Secretary for Energy, Terrorist Finance, Sanctions, and Commodities in the Bureau of Economic and Business Affairs from 2000-2003. Prior to this she was Director of the Office of the United Kingdom, Benelux, and Ireland Affairs and from 1996-1999 was Deputy Chief of Mission at the American Embassy in Kuala Lumpur, Malaysia.

Anna Borg began her Foreign Service career in 1978 after working at The World Bank. Earlier assignments have included: policy advisor to the Deputy Secretary of State (1993), policy advisor on Bosnia in the European Bureau (1992-1993), and deputy director of the Pakistan, Afghanistan, and Bangladesh Office (1990-1992). She has received a Presidential Meritorious Service Award, the 2007 Baker-Wilkins Award for DCM of the Year, the 1988 James Clement Dunn Award for FS-01 Officer of the year and State Department Superior Honor Awards.

A native of Baltimore, Maryland, she received a B.A. from Swarthmore College, M.A. from George Washington University, D.E.A. from the Ecole des Hautes Etudes en Sciences Sociales, and diploma from the National War College (1990). Her foreign languages are French and Italian.

MICHAEL BORRUS

Michael Borrus is the founding general partner of X/Seed Capital, a seed-focused early-stage venture fund that invests in entrepreneurs pursuing breakthrough innovation. Prior to founding X/Seed, he was an Executive in Residence (EIR) at Mohr Davidow Ventures (MDV) in Silicon Valley.

From 1999 to 2004, Michael led the technology banking unit at The Petkevich Group, a financial services start-up. Before that, Michael was Adjunct Professor in UC Berkeley's College of Engineering and a partner in the business consulting firm Industry and Trade Strategies. While at Berkeley, he co-founded and co-directed the Berkeley Roundtable on the International Economy.

He is the author of three books and over 70 chapters, articles and monographs on a variety of topics including management of technology, high-technology competition, international trade and investment, and financial strategies for technology companies.

Michael serves on several National Academy of Sciences/National Research Council steering committees including as Vice-Chairman of the Committee on Competing in the 21st Century: Best Practice in State and Regional Innovation Initiatives. He also serves on the board of trustees for the National Center for Women and Information Technology (NCWIT) and The UC Berkeley School of Mechanical Engineering External Advisory Board. He is a director of multiple privately held technology start-ups creating products for cleantech, life science, and information technology markets.

Michael is an honors graduate of Harvard Law School, the University of California, Berkeley, and Princeton University. He is a member of the California State Bar.

DAN BREZNITZ

Professor Dan (Danny) Breznitz (Georgia Institute of Technology, Sam Nunn School of International Affairs & The School of Public Policy, Ph.D. MIT) has extensive experience in conducting comparative in-depth research of Rapid-Innovation-Based Industries and their globalization. Dr. Breznitz's first book, Innovation and the State: Political Choice and Strategies for Growth in Israel, Taiwan, and Ireland (Yale University Press), won the 2008 Don K. Price for best book on Science and Technology given by APSA and was a finalist for the 2007 best book of the year award in political science by ForeWord Magazine. His second book (co-authored with Michael Murphree) The Run of the Red Queen: Government, Innovation, Globalization, and Economic Growth in China is forthcoming with Yale University Press in 2010. In addition, his work was published in various journals, as well as chapters in edited volumes. Breznitz is one of five young North American scholars to be selected as a 2008 Industry Study Fellow of the Sloan Foundation. Breznitz has also been an advisor on Science Technology and Innovation Policies for multinational corporations, international organizations such as the World Bank and WIPO, and local and national governments in the United States, Asia, and Europe.

During 2006 Dr. Breznitz was a visiting scholar at Stanford University's Project on Regions of Innovation and Entrepreneurship, and during 2007 he was a Visiting Fellow at the Bruegel Institute for International Economics, Brussels. His work is sponsored by the Sloan Foundation, the Kauffman Foundation, the Samuel Neaman Institute for Advance Studies, the Bi-National Science Foundation (US Israel), the NSF, Georgia Research Alliance, and the Enterprise Innovation Institute. In addition, Dr. Breznitz is the co-director with John Zysman of UC Berkeley of a collaborative study titled "Can Wealthy Nations Stay Rich in a Rapidly Changing Global Economy?" A former founder and CEO of a small software company, Dr. Breznitz is also a research affiliate of MIT's Industrial Performance Center. In addition he is a senior researcher of the Science, Technology, and Innovation Policy Program (STIP) and the academic director of the Initiative for High Tech Clusters at The Enterprise Innovation Institute (EI2), and the director of the Globalization, Innovation, and Development program at the Center for International Strategy, Technology and Policy (CISTP) in the Sam Nunn School of the Georgia Institute of Technology.

CARL DAHLMAN

Carl J. Dahlman is the Luce Professor of International Relations and Information Technology at the Edmund A. Walsh School of Foreign Service at Georgetown University. He joined Georgetown in January 2005 after more than 25 years of distinguished service at The World Bank. At Georgetown, Dr. Dahlman's research and teaching explore how rapid advances in science, technology and information are affecting the growth prospects of nations and influencing trade, investment, innovation, education and economic relations in an increasingly globalizing world. At The World Bank Dr. Dahlman served as Senior Advisor to The World Bank Institute and managed the Knowledge for Development (K4D) since 1999. Prior to that he served as staff director of the 1998-1999 World Development Report, Knowledge for Development, was the Bank's resident representative and financial sector leader in Mexico, and led divisions in the Bank's Private Sector Development, and Industry and Energy Departments. He has conducted extensive analytical work in major developing countries including Argentina, Brazil, Chile, Mexico, Russia, Turkey, India, Pakistan, China, Indonesia, Korea, Malaysia, Philippines, Thailand, and Vietnam. He has co-authored eight books on the knowledge economy in different countries and many chapters and articles education and skills, and innovation. He is currently finalizing a book on the implications of the rise of China and India for the world.

MARK DEAN

Dr. Mark E. Dean is vice president Technical Strategy and Global Operations for IBM Research. In this role, he is responsible for setting the direction of IBM's overall Research Strategy across eight worldwide labs and leading the global operations and information systems teams. An engineer by training, Dr. Dean has over 29 years with IBM, and is an IBM Fellow. He has been central to the design of a wide range of IBM products.

Dr. Dean has held various positions in several different cities and IBM divisions. Prior to his current role, he was vice president of the IBM Almaden Research Center in San Jose, California and senior location executive for Silicon Valley, overseeing more than 400 scientists and engineers doing exploratory and applied research in various hardware, software and services areas including: nanotechnology, materials science for storage systems, data management, web technologies, workplaces practices and user interfaces.

Before his appointment to the Almaden Lab in 2004, Dr. Dean was vice president for hardware and systems architecture in IBM's Systems and Technology Group in Tucson, Arizona. While there, he significantly enhanced STG's hardware and systems strategy and architectures to support continued market share growth and industry leadership in IBM's server and storage systems business. Before STG, Dr. Dean was a vice president in IBM's Storage Technology Group, focused on the company's storage systems strategy and technology roadmap.

Prior to Tucson, Dr. Dean was the VP for Systems Research at IBM's Watson Research Center in Yorktown Heights, New York, where he was responsible for the research and application of systems technologies spanning circuits to operating environments. Key technologies from his research team include petaflop supercomputer systems structures (BlueGene), digital visualization, design automation tools, Linux optimizations for servers and embedded systems, algorithms for computational science, memory compression, S/390 & PowerPC processors, embedded systems research, formal verification methods and high-speed low-power circuits.

During his career, Dr. Dean has held several engineering positions at IBM in the area of computer system hardware architecture and design in Boca Raton, Florida, Austin, Texas and Yorktown Heights, New York. He has developed all types of computer systems, from embedded systems to supercomputers, including testing of the first gigahertz CMOS microprocessor, and establishing the team that developed the Blue Gene supercomputer. He was also chief engineer for the development of the IBM PC/AT, ISA systems bus, PS/2 Model 70 & 80, the Color Graphics Adapter in the original IBM PC, and holds three of the nine patents for the original IBM PC. One invention—the Industry Standard Architecture (ISA) "bus," which permitted add-on devices like the keyboard, disk drives and printers to be connected to the motherboard—would earn election to the National Inventors Hall of Fame for Dean and colleague Dennis Moeller.

Dr. Dean received a BSEE degree from the University of Tennessee in 1979, an MSEE degree from Florida Atlantic University in 1982, and a Ph.D. in electrical engineering from Stanford University in 1992.

Dr. Dean's most recent awards include National Institute of Science Outstanding Scientist Award, member of the American Academy of Arts and Science and National Academy of Engineering, IEEE Fellow, the CCG Black Engineer of the Year, the NSBE Distinguished Engineer award, the University of Tennessee COE Dougherty Award, member of the National Inventor's Hall of Fame, and recipient of the Ronald H. Brown American Innovators Award. Dr. Dean was appointed to IBM Fellow in 1995, IBM's highest technical honor. He is a member of the IBM Academy of Technology. He has received several academic and IBM awards, including thirteen Invention Achievement Awards and six Corporate Awards. Dr. Dean has more than 40 patents or patents pending.

EUGENE HUANG

Eugene J. Huang currently serves in the White House Office of Science and Technology Policy as the senior advisor to the chief technology officer.

From August 2009 to April 2010, Mr. Huang served as the government operations director for the National Broadband Task Force at the Federal Communications Commission, and was part of the team responsible for authoring "Connecting America: The National Broadband Plan."

Mr. Huang served at the United States Department of the Treasury under two Secretaries of the Treasury from 2006 to 2009, as policy advisor to the Secretary and previously as a White House Fellow. In these roles, Mr. Huang covered a wide range of international economic and finance issues with a special responsibility for U.S. bilateral relations with China.

Previously, Mr. Huang was a visiting scholar at the Stanford Institute for Economic Policy Research (SIEPR) at Stanford University. From 2002 to 2006, Mr. Huang served the Commonwealth of Virginia under Governor Mark R. Warner as the Secretary of Technology and previously as the Deputy Secretary of Technology. At the time of his appointment as Secretary of Technology in 2004, he was the youngest cabinet member in Virginia history at the age of 28.

Mr. Huang graduated magna cum laude from the University of Pennsylvania, with a B.S. in economics from the Wharton School, a B.S. in electrical engineering, and a M.S. in telecommunications engineering. He received a Thouron Award from the University of Pennsylvania and studied at St. Peter's College, Oxford University, where he received a M.Phil., with distinction, in economic history. Mr. Huang is a term member of the Council on Foreign Relations and a member of the International Institute for Strategic Studies.

LOU JING

Lou Jing is currently serving as deputy director of the Department of Science and Technology, of the Ministry of Education

From 1998 to 2008, Ms. Lou worked in education informatization and management work, including education system infrastructure construction, resource system, systems, middleware, and user service systems; research and promotion of education informatization standardized construction work; and research of educational electronic administration construction and development; with his research receiving the National Ministry-Level Science and Technology Achievement First Class Award.

Ms. Lou participated in formulation of Phases I and II of the Education Revitalization Action Plan and the formulation work for education informatization in the education development planning of the "Twelfth Five-Year Plan."

In 2007, Ms. Lou started serving as Deputy Director of the Department of Science and Technology, mainly working in the advancement of high and new technology research and development; the construction and management of science and technology innovation and transfer platforms, such as university science and technology parks, engineering research centers, and engineering technology centers; and also researching intellectual property rights protection and organizing university science and technology strengths to benefit national innovation system construction.

Lou Jing has received bachelor's, master's, and doctoral degrees in telecommunication engineering, systems engineering, business administration, and management engineering.

KRISTINA JOHNSON

Kristina M. Johnson is currently the Under Secretary for Energy at the Department of Energy in Washington, DC. Prior to her appointment as Under Secretary, Dr. Johnson was provost and senior vice president for Academic Affairs at The Johns Hopkins University. She received her B.S. (with distinction), M.S., and Ph.D. in electrical engineering from Stanford University. After a NATO post-doctoral fellowship at Trinity College, Dublin, Ireland, she joined the University of Colorado-Boulder's faculty in 1985 as an assistant professor and was promoted to full professor in 1994. From 1994 to 1999 Dr. Johnson directed the NSF/ERC for Optoelectronics Computing Systems Center at the University of Colorado and Colorado State University, and then served as dean of the Pratt School of Engineering at Duke University from 1999 to 2007.

Dr. Johnson was named an NSF Presidential Young Investigator in 1985 and a Fulbright Faculty Scholar fellowship in 1991. Her awards include the Dennis Gabor Prize for creativity and innovation in modern optics (1993); State of Colorado and North Carolina Technology Transfer Awards (1997, 2001); induction into the Women in Technology International Hall of Fame (2003); the Society of Women Engineers Lifetime Achievement Award (2004); and in May of 2008, the John Fritz Medal, widely considered the highest award in the engineering profession. Previous recipients of the Fritz Medal include Alexander Graham Bell, Thomas Edison and Orville Wright. In December of 2009, she was awarded an honorary Doctorate of Science from the University of Alabama at Huntsville.

Dr. Johnson has 142 refereed papers and proceedings and holds 45 U.S. patents (129 U.S. and international patents) and patents pending.

A fellow of the Optical Society of America, International Electronics and Electrical Engineering (IEEE), SPIE, the International Society for

Optical Engineering (former Board Member), Dr. Johnson has served on the Board of Directors of Mineral Technologies Inc., Boston Scientific Corporation, AES Corporation and Nortel Networks. She helped found several companies, including ColorLink, Inc, SouthEast Techinventures, and Unyos.

PATRICK KEATING

Patrick Keating is vice president, China 21C Leadership, and Cisco Managing Director, Guanghua Leadership Institute in charge of leadership programs for Chinese government officials and enterprise executives. Pat co-leads Cisco's initiative to build a Leadership Institute in strategic partnership with the Guanghua School of Management at Peking University. In his previous role, Pat was responsible for worldwide leadership and executive education programs at Cisco. Pat has held positions in government, industry, and academia spanning the areas corporate transformation, financial management, and information technology. Pat holds a Ph.D. from Penn State University where he also earned B.S. in electrical engineering. Pat holds a master's degree in public policy from the University of Michigan. Prior to Cisco, Pat was professor of business administration at San Jose State University.

GINGER LEW

Ms. Lew is senior advisor to the White House National Economic Council and the SBA Administrator. She provides economic policy advice on a broad range of matters that impact small businesses. In addition, she co-chairs the White House Interagency Group on Innovation and Entrepreneurship.

Prior to joining the Obama Administration, Ms. Lew was the managing partner of a communications venture capital fund, and a venture advisor to a Web 2.0 venture fund.

Under the Clinton Administration, Ms. Lew was the deputy administrator and chief operating officer of the U.S. Small Business Administration where she provided day to day management and operational oversight of a $42 billion loan portfolio. Before joining SBA, Ms. Lew was the General Counsel at the U.S. Department of Commerce where she specialized in international trade issues. Ms. Lew was unanimously confirmed by the United States Senate for both positions.

For the past ten years, Ms. Lew was Chairman and board member of an investment fund based in Europe. She has served on the boards of publicly traded companies, private companies, and nonprofit organizations.

C.D. MOTE, JR.

In September 1998, C. D. (Dan) Mote, Jr. began his tenure as president of the University of Maryland and Glenn L. Martin Institute Professor of Engineering. He was recruited to lead the University of Maryland to national eminence under a mandate by the state. Since assuming the presidency, he has encouraged an environment of excellence across the University and given new impetus to the momentum generated by a talented faculty and student body. Under his leadership, academic programs have flourished. In 2005, the University was ranked 18th among public research universities, up from 30th in 1998. President Mote has emphasized broad access to the university's model, enriched undergraduate curriculum programs and launched the Baltimore Incentive Awards Program to recruit and provide full support to high school students of outstanding potential who have overcome extraordinary adversity during their lives.

President Mote has spurred the university to lead the state in the development of its high-tech economy, especially in the information and communication, bioscience and biotechnology, and nano-technology sectors. President Mote has greatly expanded the university's partnerships with corporate and federal laboratories and successfully negotiated to bring to the College Park area the first Science Research Park sponsored by the People's Republic of China. Under his leadership, the University has established a research park, The University of Maryland Enterprise Campus, M-Square, located on a 115-acre site adjacent to the University of Maryland/College Park Metro with 3 million square feet of development potential. Among its first tenants are the Center for Advanced Study of Language, a joint venture of the University and Department of Defense, and the National Oceanic and Atmospheric Administration's new World Weather and Climate Prediction Center.

During President Mote's second year in office, the University began the largest building boom in its history, with more than $100 million in new projects breaking ground that year. New facilities address every aspect of university life, from the arts to recreation to classrooms and laboratories, and, in creative partnership with the private sector, new residential facilities. Highlights of the construction activity include the stunning Clarice Smith Performing Arts Center; the Comcast Center, a state of the art sports complex; a high-tech research greenhouse; and new classrooms for chemistry, computer science, business and engineering. President Mote also led the development of a new Facilities Master Plan for development in the next 20 years, which is noted for its emphasis on environmental stewardship.

Dr. Mote is a leader in the national dialogue on higher education and his analyses of shifting funding models have been featured in local and national media. He has testified on major educational issues before Congress, representing the University and higher education associations on the problem of visa barriers for international students and scholars and on deemed export control issues. He has been asked to serve on a high-level National Academies Committee appointed at the request of the Senate Energy Subcommittee of the Senate Energy and Natural Resources Committee to identify challenges to United States leadership in key areas of science and technology and to be a member of the Leadership Council of the National Innovation Initiative, an activity of the Council on Competitiveness. He has served as vice chair of the Department of Defense Basic Research Committee, and is a member of the Council of the National Academy of Engineering. In 2004-2005, he served as President of the Atlantic Coast Conference. In its last ranking in 2002, Washington Business Forward magazine counted him among the top 20 most influential leaders in the region.

Prior to assuming the presidency at Maryland, Dr. Mote served on the University of California, Berkeley faculty for 31 years. From 1991 to 1998, he was vice chancellor at Berkeley, held an endowed chair in mechanical systems and was president of the UC Berkeley Foundation. He led a comprehensive capital campaign for Berkeley that raised $1.4 billion. He earlier served as chair of Berkeley's Department of Mechanical Engineering and led the department to its number one ranking in the National Research Council review of graduate program effectiveness.

Dr. Mote's research lies in dynamic systems and biomechanics. Internationally recognized for his research on the dynamics of gyroscopic systems and the biomechanics of snow skiing, he has produced more than 300 publications, holds patents in the United States, Norway, Finland and Sweden, and has mentored 56 Ph.D. students. He received the B.S., M.S. and Ph.D. in mechanical engineering from the University of California, Berkeley. President Mote has received numerous awards and honors, including the Humboldt Prize awarded by the Federal Republic of Germany. He is a recipient of the Berkeley Citation, an award from the University of California-Berkeley similar to the honorary doctorate, and was named Distinguished Engineering Alumnus. He has received two honorary doctorates. He is a member of the U.S. National Academy of Engineering and serves on its Council, and is a member of the American Academy of Arts and Sciences. He was elected to Honorary Membership in the ASME International, its most distinguished recognition, and is a Fellow of the International Academy of Wood Science, the Acoustical Society of America, and the American Association for the Advancement of Science. In Spring 2005, he was named recipient of the 2005 J. P. Den Hartog award by the ASME International Technical Committee on Vibration and Sound to honor his lifelong contribution to the teaching and/or practice of vibration engineering. In Fall 2005, he received the 2005 Founders Award from the National Academy of Engineering in recognition of his comprehensive body of work on the dynamics of moving flexible structures and for leadership in academia.

ROBIN NEWMARK

Robin L. Newmark is director of the Strategic Energy Analysis Center at the National Renewable Energy Laboratory (NREL). Prior to joining NREL, Dr. Newmark was at the Lawrence Livermore National Laboratory (LLNL), where her research focused primarily on energy, environment and national security. In recent years, she has led or contributed to programs involving energy, climate and water issues, including the interdependence of water and energy systems, including a water initiative with components addressing the impacts of climate change on water resources, assessing denitrification in agricultural regions, and the development of energy-efficient, selective water treatment technologies. Dr. Newmark is an active member of the multi-national laboratory Energy-Water Nexus working group, the World Resources Institute Carbon Capture and Sequestration (CCS) Stakeholder Group and the U.S.-China Expert CCS Steering Committee. She is an author of over 50 papers, reports and patents, past vice president of the Near Surface Geophysics Section of the Society of Exploration Geophysicsts, past Associate Editor for Geophysics, and a Fellow of both the Renewable and Sustainable Energy Institute at the University of Colorado, Boulder and the Center of Integrated Water Research at the University of California at Santa Cruz.

Dr. Newmark holds a B.S. from the Massachusetts Institute of Technology, where she was selected Phi Beta Kappa, a M.S. from the University of California at Santa Cruz, an M.Phil and a Ph.D from Columbia University.

CHARLES VEST

Charles M. Vest is president of the U.S. National Academy of Engineering and president emeritus of the Massachusetts Institute of Technology. A professor of mechanical engineering at MIT and formerly at the University of Michigan, he served on the U.S. President's Council of Advisors on Science and Technology from 1994 to 2008, and chaired the President's Committee on the Redesign of the Space Station and the Secretary of Energy's Task force on the Future of Science at DoE. He was a member of the Commission on the Intelligence Capabilities of the United States Regarding Weapons of Mass Destruction and the Secretary of Education's Commission on the Future of Higher Education. He was vice chair of the U.S. Council on Competitiveness for seven years, has served on the boards of DuPont and IBM, and was awarded the 2006 National Medal of Technology. He is the author of a book on holographic interferometry and two books on higher education. Constant themes throughout his career have included the quality and diversity of the U.S. engineering workforce; sustained excellence of U.S. higher education; global openness to the flow of people, education, and ideas; university-government-industry partnership; and the innovative capacity of the United States.

Dr. Vest holds ten honorary doctorates and received the 2006 National Medal of Technology.

REN WEIMIN

Ren Weimin is currently serving as deputy director of the Academy of Macroeconomic Research at the National Development and Reform Commission.

From 2003 to 2009, Mr. Ren served as deputy inspector in the Office of the National Development and Reform Commission.

From 1998 to 2003, Mr. Ren served as director in the Office of Economic Restructuring, State Council; and deputy director of the Department of Secretarial and Administrative Affairs.

From 1994 to 1998, Mr. Ren served as deputy director and duty office director, in the Department of Training, at the Office of the Commission for Economic Restructuring.

Mr. Ren has worked for many years for state agencies in China in cadre training and administrative work.

CHARLES WESSNER

Charles Wessner is a National Academy Scholar and director of the Program on Technology, Innovation, and Entrepreneurship. He is recognized nationally and internationally for his expertise on innovation policy, including public-private partnerships, entrepreneurship, early-stage financing for new firms, and the special needs and benefits of high-technology industry. He testifies to the U.S. Congress and major national commissions, advises agencies of the U.S. government and international organizations, and lectures at major universities in the United States and abroad. Reflecting the strong global interest in innovation, he is frequently asked to address issues of shared policy interest with foreign governments, universities, research institutes, and international organizations, often briefing government ministers and senior officials. He has a strong commitment to international cooperation, reflected in his work with a wide variety of countries around the world.

Dr. Wessner's work addresses the linkages between science-based economic growth, entrepreneurship, new technology development, university-industry clusters, regional development, small-firm finance and public-private partnerships. His program at the National Academies also addresses policy issues associated with international technology cooperation, investment, and trade in high-technology industries.

Currently, he directs a series of studies centered on government measures to encourage entrepreneurship and support the development of new technologies and the cooperation between industry, universities, laboratories, and government to capitalize on a nation's investment in research. Foremost among these is a congressionally mandated study of the Small Business Innovation Research (SBIR) Program, reviewing the operation and achievements of this $2.3 billion award program for small companies and start-ups. He is also directing a major study on best practice in regional innovation programs, entitled Competing in the 21st Century: Best Practice in State & Regional Innovation Initiatives. Today's meeting on "Building the 21st Century: U.S.-China Cooperation on Science, Technology, and Innovation," forms part of a complementary, global analysis entitled Comparative Innovation Policy: Best Practice in National Technology Programs. The overarching goal of Dr. Wessner's work is to develop a better understanding of how we can bring new technologies forward to address global challenges in health, climate, energy, water, infrastructure, and security.

ALAN WM. WOLFF

Alan Wm. Wolff holds the position of distinguished research professor, Graduate School of International Policy, at the Monterey Institute of International Studies. He also serves as Of Counsel at the international law firm of Dewey & LeBoeuf and leads the firm's International Trade Practice. He served as United States Deputy Special Representative for Trade Negotiations (1977-1979) in the Carter Administration, holding the rank of ambassador, after having served as General Counsel of the agency from 1974 to 1977. As Deputy Trade Representative, he played a key role in the formulation of American trade policy and its implementation. From 1968 to 1973, he was an attorney dealing with international monetary, trade, and development issues at the Treasury Department.

Ambassador Wolff is a member of the National Academies' Board on Science, Technology, and Economic Policy (STEP Board) from 1997 to present. He is a lifetime "National Associate" of the National Academies. Ambassador Wolff chairs the Academies' Committee on Comparative Innovation Policy: Best Practice in National Technology Programs. Ambassador Wolff is Chairman of the Advisory Board of the Institute for Trade and Commercial Diplomacy; and is a member of the U.S. Department of State's Advisory Committee on International Economic Policy; the Advisory Committee of the Peterson Institute for International Economics; the Board of National Foreign Trade Council (NFTC); the United States Council for International Business; the Council on Foreign Relations, and the American Society of International Law. He is also a Board Member of the U.S.-China Legal Cooperation Fund and of the National Trade Council Foundation. He served on the Board of Trustees of Monterey Institute for International Studies from 1992 to 2001.

Ambassador Wolff is recognized in Chambers USA - America's Leading Lawyers for Business as a leader in the field of International Trade and is recognized in Best Lawyers in America as a leader in the field of International Trade and Finance Law.

Ambassador Wolff has co-authored books and published numerous papers on trade and U.S. trade law. He received his Juris Doctor degree from Columbia University and his B.A. from Harvard College.

YANG XIANWU

Yang Xianwu is currently serving as deputy director, Department of High Technology Development and Commercialization, at the Ministry of Science and Technology

Joining the Ministry in 1986, Mr. Yang has worked in areas of science and technology planning, reform and restructuring, high-tech know-how transfer and commercialization. He took part in drafting China's 9th, 10th, and 11th five-year national science and technology programs.

Since 1998, Mr. Yang has been dedicated to high-tech commercialization, including development of national high-tech industry zones, high-tech business incubators, university high-tech parks, and center of productivity boosting.

Mr. Yang is responsible for advancing R&D and commercialization of information technology and space technology.

CHEN YING

Chen Ying is currently serving as the deputy director of the Department of Software Service Industry of the Ministry of Industry and Information Technology. Chen Ying started working in industry administration and policy research, enactment, and implementation in 1995. He has participated in the drafting and implementation stages of China's most important software industry policies, such as encouraging software and integrated circuit industry development, promoting the Chinese software industry's recent fast development, promoting Chinese software intellectual property right protection work, and promoting, organizing, and implementing the pre-installation of official operating systems in computers sold in China before leaving the factory.

C

PARTICIPANTS LIST

Anna Barker
National Cancer Institute

Anna Borg
U.S. Department of State

Paul Beaton
Yale University

Chen Biao
Shenzhen DRC

Xu Bin
Hi-tech Department, NDRC

Robert Boege
ASTRA

Bill Bonvillian
Massachusetts Institute of
Technology

Michael Borrus
X/Seed Capital

Dan Breznitz
Georgia Institute of Technology

Steve Campbell
National Institute of Standards
and Technology

Agnes Chang
Interpreter

Yuyu Chen
Peking University

McAlister Clabaugh
The National AcademiesTara
Collison
Cisco Systems, Inc.

Carl Dahlman
Georgetown University

Vittorio Daniore
Embassy of Italy

Mark Dean
IBM

Ding Di
Chongqing DRC

David Dierksheide
The National Academies

Lucas DiLeo
Broadland Advisors, LLC

Zhang Dingyu
Bureau of Land and Resources
and House Management

Kai Duh
University of Maryland

Bing Wei Edwards
Cisco Systems, Inc.

Will Edwards
U.S. Department of Energy

Steven Ezell
Information Technology and
Innovation Foundation

Nick Fetchko
Telecommunications Industry
Association

Becky Fraser
U.S. Chamber of Commerce

Adam Gertz
The National Academies

Kenneth Gertz
University of Maryland

Fang Haiyang
The People's Government of
ShaPingBa District of
Chongqing

He Hao
Chengdu Shuangliuxian
Committee

Kathryn Holmes
ASME

Robert Hormats
U.S. Department of State

John Horrigan
U.S. Federal Communications
Commission

Angel Hsu
Yale University

Mengfei Huang
The National Academies

EugeneHuang
*White House Office of Science
and Technology Policy*

Mu Huaping
Chongqing Economic &
Information Committee

Jim Hurd
GreenScience Exchange

Takashi Inutsuka
Embassy of Japan

Kenan Jarboe
Athena Alliance

Cheng Jianlin
Comprehensive Economy
Department, NDRC

Lou Jing
*Science & Technology
Department, Ministry of
Education*

Xu Jing
Tariff Policy Department
Ministry of Finance

Kristina Johnson
U.S. Department of Energy

Burk Kalweit
ASTRA

PatrickKeating
Cisco Systems, Inc.

Taffy Kingscott
IBM

Ginger Lew
White House National
Economic Council

Dapeng Lin
Cisco Systems, Inc.

Joseph Lin
Cisco Systems, Inc.

Ying Lowrey
U.S. Small Business
Administration

Philipp Marxgut
Embassy of Austria

Stephen Merrill
The National Academies

Brad Miller
American Chemical Society

Li Min Xue
Chengdu Hitech Zone
Committee

C.D. "Dan" Mote
University of Maryland

Robin Newmark
National Renewable Energy
Laboratory

Brad Peganoff
RTI International

Mark Peshoff
Cisco Systems, Inc.

Anne Pizatto
Telecommunications Industry
Association

Zhou Quanhong
Mriam Quintal
Lewis-Burke Associates

Jeannine Ray
Cisco Systems, Inc.

Paul Ross
Alcatel-Lucent Bell Labs

Phil Rudd
University of Michigan

Larry Schuette
Office of Naval Research

Juan Serrano
Embassy of Spain

RD Shelton
WTEC

Sujai Shivakumar
The National Academies

Marc Stanley
National Institute of Standards
and Technology

Calvin Sui
Cisco Systems, Inc.

Makito Takami
NEDO

Bill Taylor
SEMATECH

Ted Theofrastous
Nortech Energy Enterprise

Albert Ting
U.S. Department of Commerce

Richard Van Noorden
Nature

Jason Van Wey
Washington University in St.
Louis

Charles Vest
National Academy of
Engineering

Derek Vollmer
The National Academies

Cyrus Wadia
White House Office of Science
and Technology Policy

Wang Wei
Commodity & Labor Tax
Department SAT

Ren Weimin
Academy of Macroeconomic
Research NDRC

Wang Wenhua
Foreign Affairs Department,
NDRC

Charles Wessner
The National Academies

Zach Wilson
CM2

Lorel Wisniewski
National Institute of Standards
and Technology

Alan Wm. Wolff
Dewey & LeBoeuf, LLP

Julian Wong
Center for American Progress

Irene Wu
U.S. Federal Communications
Commission

Yang Xianwu
High & New Technology
Department, Ministry of Science
& Technology, China

Ziyun Xu
Interpreter

Karl Xuan
Cisco Systems, Inc.

Li Min Xue
Chengdu Hi-Tech Zone

Wang Xue
Sichuan Department of Health

Ophelia Yeung
SRI International

Bin Yin
Cisco Systems, Inc.

Bo-WenYin
Cisco Systems, Inc.

Chen Ying
Software Department, Ministry
of Industry and Information
Technology

Zhu Yingjuan
Economy & Trade Department,
NDRC

Sui Yun
Interpreter

Huang Zhengzhang
Sichuan DRC

John Zysman
Berkley Roundtable on the
International Economy

D

BIBLIOGRAPHY

Acs, Zoltan, and David Audretsch. 1990. *Innovation and Small Firms.* Cambridge, MA: The MIT Press.

Aerts, Kris, and Dirk Czarnitzki. 2005. "Using Innovation Survey Data to Evaluate R&D Policy: The Case of Flanders." Catholic University of Leuven, Department of Applied Economics and Steunpunt O&O Statistieken.

Aerts, Kris, and Dirk Czarnitski. 2006. "The Impact of Public R&D Funding in Flanders." *IWT Studies* 54.

Aghion, Phillipe, Robin Burgess, Stephen Redding, and Fabrizio Zilibotti. 2003. "The Unequal Effects of Liberalization: Theory and Evidence from India." Washington, DC: Center for Economic Policy Research. Agrawal, A. and R. Henderson. 2002. "Putting Patents in Context: Exploring Knowledge Transfer from MIT." *Management Science* 48(1):44-60.

Ahluwalia, Montek Singh. 2001. "State Level Reforms Under Economic Reforms in India." Stanford University Working Paper No. 96, March.

Aizcorbe, A., K. Flamm, and A. Kurshid. 2002. "The Role of Semiconductor Inputs in IT Hardware Price Decline: Computers vs. Communications." Federal Reserve Finance and Economics Discussion Paper 2002-37. Washington, DC: The Federal Reserve Board of Governors. August 2002; revised 2004.

Aizcorbe, A., S. Oliner, and D. Sichel. 2006. "Shifting Trends in Semiconductor Prices and the Pace of Technological Progress." Federal Reserve Board Finance and Economics Discussion Series Working Paper No. 2006-44.September.

Alic, John A., Lewis M. Branscomb, Harvey Brooks, Ashton B. Carter, and Gerald L. Epstein. 1992. Beyond Spin-off: Military and Commercial Technologies in a Changing World. Boston, MA: Harvard Business School Press.

Allen, John. 2011. "House continuing resolution would bar NASA from China ties." *Politico.* February 12.

Allen, Stuart D., Albert N. Link, and Dan T. Rosenbaum. 2007. "Entrepreneurship and Human Capital: Evidence of Patenting Activity from the Academic Sector." *Entrepreneurship Theory and Practice* 31(6):937-951.

Allison, J., and M. Lemley. 1998. "Empirical Evidence on the Validity of Litigated Patents." *AIPLA Quarterly Journal* 26:185-277.

Altenburg, Tilman, Hubert Schmitz, and Andreas Stamm. 2008. "Breakthrough: China's and India's Transition from Production to Innovation." *World Development* 36(2):325-344.

Amsden, Alice H. 2001. *The Rise of "the Rest": Challenges to the West from Late-industrializing Economies.* Oxford, UK: Oxford University Press.

Amsden, Alice H. and Wan-wen Chu. 2003. *Beyond Late Development: Taiwan's Upgrading Policies.* Cambridge, MA: The MIT Press.

Amsden, Alice H., Ted Tschang, and Akira Goto. 2001. "Do Foreign Companies Conduct R&D in Developing Countries?" Tokyo, Japan: ADB Institute.

Aoki, Reiko, and Sadao Nagaoka. 2004. "The Consortium Standard and Patent Pools." *The Economic Review (Keizai Kenkyu)* 55(4):345-356.

Aoki, Reiko, and Sadao Nagaoka. 2005. "Coalition Formation for a Consortium Standard through a Standard Body and a Patent Pool: Theory and Evidence from MPEG2, DVD and 3G." IIR Working Paper WP#05-01. February.

Applied Research Institute, Inc. 2006. Survey of the Environment for Startups. Applied Research Institute. November.

Archibugi, Danielle, Jeremy Howells, and Jonathan Michie, eds. 1999. *Innovation Policy and the Global Economy.* Cambridge, UK: Cambridge University Press.

Argyres, N. S., and J. P. Liebeskind. 1998. "Privatizing the Intellectual Commons: Universities and the Commercialization of Biotechnology." *Journal of Economic Behavior & Organization* 35:427-454.

Arora, A., and R. P. Merges. 2004. "Specialized Supply Firms, Property Rights, and Firm Boundaries." *Industrial and Corporate Change* 13(3):451-476.

Arora, A., M. Ceccagnoli, and W. Cohen. 2001. "R&D and the Patent Premium." Carnegie-Mellon University and INSEAD: paper presented at the ASSA Annual Meetings. January 2002. Atlanta, Georgia.

Arora, A., A. Fosfuri, and A. Gambardella. 2003. "Markets for Technology and Corporate Strategy." In O. Granstrand, ed., *Economics, Law, and Intellectual Property*, Boston, MA: Kluwer Academic Publishers.

Arrow K. J. 2000. "Increasing Returns: Historiographic Issues and Path Dependence." *European Journal of the History of Economic Thought* 7(2):171-180.

Arrow, Kenneth. 1962. The Rate and Direction of Inventive Activity. Princeton, NJ: Princeton University Press. Pp. 609-626.

Arthur, W. 1989. "Competing Technologies, Increasing Returns, and Lock-in by Historical Small Events." *Economic Journal* 99(2):116-131.

Asheim, Bjorn T. et al., eds. 2003. *Regional Innovation Policy for Small-medium Enterprises*. Cheltenham, UK: Edward Elgar.

Association of University Research Parks. 1998. "Worldwide Research & Science Park Directory 1998." New York: BPI Communications.

Association of University Research Parks. 2008 "The Power of Place: A National Strategy for Building America's Communities of Innovation." Tucson, AZ: Association of University Research Parks.

Athreye, Suma S. 2000. "Technology Policy and Innovation: The Role of Competition Between Firms." In Pedro Conceicao et al., eds. *Science, Technology, and Innovation Policy: Opportunities and Challenges for the Knowledge Economy*. Westport, CT: Quorum Books.

Atkinson, Robert. 2004. *The Past and Future of America's Economy-Long Waves of Innovation that Power Cycles of Growth*. Cheltenham, UK: Edward Elgar.

Atkinson, Robert. 2006. "Is the Next Economy Taking Shape?" *Issues in Science and Technology*. Winter.

Audretsch, D. B., ed. 1998. *Industrial Policy and Competitive Advantage*, Volumes 1 and 2. Cheltenham, UK: Edward Elgar.

Audretsch, D. B. 1998. "Agglomeration and the Location of Innovative Activity." *Oxford Review of Economic Policy* 14(2):18-29.

Audretsch, D. B. 2001. "The Prospects for a Technology Park at Ames: A New Economy Model for Industry-Government Partnership?" In National Research Council. *A Review of the New Initiatives at the NASA Ames Research Center*. Charles W. Wessner, ed. Washington, DC: National Academy Press.

Audretsch, D. B. 2006. *The Entrepreneurial Society*. Oxford, UK: Oxford University Press.

Audretsch, D. B., and M. P. Feldman. 1996. "R&D Spillovers and the Geography of Innovation and Production." *American Economic Review* 86(3):630-640.

Audretsch, D. B., and M. P. Feldman. 1999. "Innovation in Cities: Science-based Diversity, Specialization, and Localized Competition." *European Economic Review* 43(2):409-429.

Audretsch, D. B., B. Bozeman, K. L. Combs, M. P. Feldman, A. N. Link, D. S. Siegel, P. Stephan, G. Tassey, and C. Wessner. 2002. "The Economics of Science and Technology." *Journal of Technology Transfer* 27:155-203.

Audretsch, D. B., H. Grimm, and C. W. Wessner. 2005. *Local Heroes in the Global Village: Globalization and the New Entrepreneurship Policies*. New York: Springer.

Auerswald, Philip E., Lewis M. Branscomb, Nicholas Demos, and Brian K. Min. 2005. *Understanding Private-Sector Decision Making for Early-Stage Technology Development: A "Between Invention and Innovation Project" Report*. NIST GCR 02-841A. Gaithersburg, MD: National Institute of Standards and Technology.

Augustine, Norman. 2007. *Is America Falling Off the Flat Earth?* Washington, DC: The National Academies Press.

Baker, Stephen. 2005. "New York's Big Hopes for Nano." *BusinessWeek* February 4.

Bakouros, Y. L., D. C. Mardas, and N. C. Varsakelis. 2002. "Science Parks, a High-Tech Fantasy? An Analysis of the Science Parks of Greece." *Technovation* 22(2):123-128.

Baldwin, J. R., P. Hanl, and D. Sabourin. 2000. "Determinants of Innovative Activity in Canadian Manufacturing Firms: The Role of Intellectual Property Rights." Statistics Canada Working Paper No. 122. March.

Baldwin, John Russel, and Peter Hanel. 2003. *Innovation and Knowledge Creation in an Open Economy: Canadian Industry and International Implications*. Cambridge, UK: Cambridge University Press.

Balzat, Markus, and Andreas Pyka. 2006. "Mapping National Innovation Systems in the OECD Area." *International Journal of Technology and Globalisation* 2(1-2):158-176.

Baptista, R. 1998. "Clusters, Innovation, and Growth: A Survey of the Literature." In G. M. P. Swann, M. Prevezer, and D. Stout, eds. *The Dynamics of Industrial Clustering*. Oxford, UK: Oxford University Press.

Bartzokas, Anthony, and Morris Teubal. 2002. "The Political Economy of Innovation Policy Implementation in Developing Countries." *Economics of Innovation and New Technology* 11(4-5).

Battelle. 2009. *R&D Magazine* December.

Beinhocker, Eric D. 2007. *Origin of Wealth—Evolution, Complexity, and the Radical Remaking of Economics.*Cambridge, MA: Harvard Business School.

Bennis, Warren, and Patricia Ward Biederman. 1997. *Organizing Genius*. New York: Basic Books.

Berlin, Leslie. *The Man Behind the Microchip: Robert Noyce and the Invention of Silicon Valley*. 2005. New York: Oxford University Press.

Bessen, J., and M. J. Meurer. 2005. "The Patent Litigation Explosion." Research on Innovation and Boston University School of Law: manuscript. August.

Bhidé, Amar. 2006. "Venturesome Consumption, Innovation and Globalization." Paper presented at the Centre on Capitalism & Society and CESifo Venice Summer Institute 2006, "Perspectives on the Performance of the Continent's Economies," 21-22 July 2006. Held at Venice International University. San Servolo, Italy.

Biegelbauer, Peter S., and Susana Borras, eds. 2003. *Innovation Policies in Europe and the U.S.: The New Agenda*. Aldershot, UK: Ashgate.

Birch, David. 1981. "Who Creates Jobs?" *The Public Interest* 65:3-14.

Blanpied, William A. 1998. "Inventing U.S. Science Policy." *Physics Today* 51(2):34-40.

Block, Fred, and Matthew Keller. 2008. "Where Do Innovations Come From? Transformations in the U.S. National Innovation System, 1970-2006." The Information Technology and Innovation Forum. July.

Blomström, Magnus, Ari Kokko, and Fredrik Sjöholm. 2002. "Growth & Innovation Policies for a Knowledge Economy: Experiences from Finland, Sweden, & Singapore." EIJS Working Paper, Series No. 156.

Bonvillian, William B. 2006. "Power Play, The DARPA Model and U.S. Energy Policy." *The American Interest* II(2):39-48.

Borras, Susana. 2003. *The Innovation Policy of the European Union: From Government to Governance*. Cheltenham, UK: Edward Elgar.

Borrus, Michael, and Jay Stowsky. 2000. "Technology Policy and Economic Growth." In Charles Edquist and Maureen McKelvey, eds. *Systems of Innovation: Growth, Competitiveness and Employment*, Vol. 2. Cheltenham, UK: Edward Elgar.

Bradsher, Pete. 2009. "China-U.S. Trade Dispute Has Broad Implications." *New York Times* September 14.

Branscomb, L. M., and P. E. Auerswald. 2001. *Taking Technical Risks: How Innovators, Executives, and Investors Manage High-Tech Risks.* Boston, MA: The MIT Press.

Branscomb, Lewis M., and Philip E. Auerswald. 2002. *Between Invention and Innovation: An Analysis of Funding for Early-Stage Technology Development.* NIST GCR 02-841. Gathersburg, MD: National Institute of Standards and Technology. November.

Braudel, Fernand. 1973. *Capitalism and Material Life 1400-1800.* London, UK: Harper Colophon Books.

Breschi, S. and F. Lissoin. 2001. "Knowledge Spillovers and Local Innovation Systems: A Critical Survey." *Industrial and Corporate Change* 10(4):975-1005.

Breznitz, Dan. 2007. *Innovation in the State: Political Choice and Strategies for Growth in Israel, Taiwan, and Ireland.* New Haven, CT: Yale University Press.

Breznitz, D. and Murphree. 2011. *Run of the Red Queen; Government, Innovation, and Globalization and Economic Growth in China.* New Haven, CT: Yale University Press.

Broder, J. and J. Ansfield. 2009 "China and U.S. Seek a Truce on Greenhouse Gases." *The New York Times. June 7*

Browning, L., and J. Shetler. 2000. *SEMATECH: Saving the U.S. Semiconductor Industry.* College Station, TX: Texas A&M University Press.

Bush, Nathan. 2005. "Chinese Competition Policy, It Takes More than a Law." *China Business Review.* May-June.

Bush, Vannevar. 1945. *Science: The Endless Frontier.* Washington, DC: U.S. Government Printing Office.

Capron, Henri, and Michele Cincera. 2006. *Strengths and Weaknesses of the Flemish Innovation System: An External Viewpoint.* Brussels, Belgium: IWT.

Caracostas, Paraskevas, and Ugur Muldur. 2001. "The Emergence of the New European Union Research and Innovation Policy." In P. Laredo and P. Mustar, eds. *Research and Innovation Policies in the New Global Economy: An International Comparative Analysis.* Cheltenham, UK: Edward Elgar.

Castells, M., and P. Hall. 1994. *Technopoles of the World.* London, UK: Oxford University Press.

Cebrowski, Arthur, and John Garska. 1998. "Network Centric Warfare: Its Origin and Future." U.S. Naval Institute Proceedings. January.

Center for Economic Development and Business Research. 2008.
"Kansas Aviation Manufacturing,"Wichita State University: W.
Frank Barton School of Business. September

Chambers, John, ed. 1999. *The Oxford Companion to American Military
History*. Oxford, UK: Oxford University Press.

Chan, K. F., and Theresa Lau. 2005. "Assessing Technology Incubator
Programs in the Science Park: The Good, the Bad and the Ugly."
Technovation 25(10):1215-1228.

Chand, Satish, and Kunal Sen. 2002. "Trade Liberalization and
Productivity Growth: Evidence from Indian Manufacturing." *Review
of Development Economics* 6, February.

Chang, Connie, Stephanie Shipp, and Andrew Wang. 2002. "The
Advanced Technology Program: A Public-Private Partnership for
Early-stage Technology Development." *Venture Capital* 4(4): 363-
370.

Chao, Loretta. 2011. "China Plans to Ease Rules That Irked Companies."
The Wall Street Journal July 1.

Chesbrough, Henry. 2003. *Open Innovation: The New Imperative for
Creating and Profiting from Technology*. Cambridge, MA: Harvard
Business School Press. April.

China Daily. 2008. "China Luring 'Sea Turtles' Home." December 18.

China Trade Extra. 2005. China Agrees to Delay Software Procurement
Rule While Talking with U.S. July 11.

Chinese Ministry of Finance. 2006. Opinions of the Ministry of Finance
on Implementing Government Procurement Policies That Encourage
Indigenous Innovation. Cai Ku [2006] No. 47, June 13.

Chordà, I. M. 1996. "Towards the Maturity State: An Insight into the
Performance of French Technopoles." *Technovation* 16(3):143-152.

Chu Ng, Ying and Sung-ko Li, 2009. "Efficiency and productivity
growth in Chinese universities during the post-reform period." *China
Economic Review:* Volume 20

Chuma, Hiroyuki. 2006. "Increasing Complexity and Limits of
Organization in the Microlithography Industry: Implications for
Science-based Industries." *Research Policy* 35:394-411.

Chuma, Hiroyuki, and Norikazu Hashimoto. 2007. "Moore's Law,
Increasing Complexity and Limits of Organization: Modern
Significance of Japanese DRAM ERA." NISTEP Discussion Paper
No. 44. National Institute of Science and Technology Policy.

Cimoli, Mario, and Marina della Giusta. 2000. "The Nature of
Technological Change and its Main Implications on National and
Local Systems of Innovation." IIASA Interim Report IR-98-029.

Cincera, Michele. 2006. Comparison of Regional Approaches to Foster Innovation in the European Union: The Case of Flanders. Brussels, Belgium: IWT.

Cincera, Michele. 2006. R&D Activities of Flemish Companies in the Private Sector: An Analysis for the Period 1998-2002. Brussels, Belgium: IWT.

Clark, B. 1995. Places of Inquiry. Berkeley, CA: University of California Press.

Clarke, P. 2004. "LETI, Crolles Alliance Open $350-million 32-nm Research Fab." *EE Times* April 24.

Clarke, P., M. LaPedus, and M. Santarini. 2005. "IBM-led Consortium to Build Fab in N.Y." *EE Times* January 5.

Clough, G. Wayne. 2007. "The Role of the Research University in Fostering Innovation." The Americas Competitiveness Forum. June 12.

Coakes, Elayne, and Peter Smith. 2007. "Developing Communities of Innovation by Identifying Innovation Champions." *The Learning Organization: An International Journal* 14(1):74-85.

Coburn, Christopher, and Dan Berglund. 1995. *Partnerships: A Compendium of State and Federal Cooperative Programs*. Columbus, OH: Battelle Press.

Cohen, W. 2002. "Thoughts and Questions on Science Parks." Presented at the National Science Foundation Science Parks Indicators Workshop. University of North Carolina at Greensboro.

Cohen, W., R. Florida, and R. Goe. 1992. *University-Industry Research Centers in the United States*. Pittsburgh, PA: Carnegie-Mellon University.

Cohen, W., R. Nelson, and J. Walsh. 2000. "Protecting Their Intellectual Assets: Appropriability Conditions and Why U.S. Manufacturing Firms Patent (or Not)." NBER Working Paper 7552. Cambridge, MA: National Bureau of Economic Research.

Cohen, W. M., A. Goto, A. Nagata, R. R. Nelson, and J. P. Walsh. 2002. "R&D Spillovers, Patents and the Incentives to Innovate in Japan and the United States." *Research Policy* 1425:1-19.

Combs, Kathryn L., and Albert N. Link. 2003. "Innovation Policy in Search of an Economic Paradigm: The Case of Research Partnerships in the United States." *Technology Analysis & Strategic Management* 15(2).

Conant, Jennet. 2002. *Tuxedo Park*. New York: Simon & Shuster.

Conant, Jennet. 2005. *109 East Palace*. New York: Simon & Shuster.

Cooper, Helene and M. Lander. 2011. "U.S. Shifts Focus to Press China for Market Access." *New York Times.* January 18

Council on Competitiveness. 2005. *Innovate America: Thriving in a World of Challenge and Change.* Washington, DC: Council on Competitiveness.

Council of Economic Advisors. 1995. Economic Report to the President.

Council on Government Relations. 2000. *Technology Transfer in U.S. Research Universities: Dispelling Common Myths.* Washington, DC: Council on Government Relations.

Crafts, N. F. R. 1995. "The Golden Age of Economic Growth in Western Europe, 1950-1973." *Economic History Review* 3:429-447.

Czarnitzki, Dirk, and Niall O'Byrnes. 2007. "Innovation and the Impact on Productivity in Flanders." *Tijdschrift voor Economie en Management* 52(2).

Dahlman, Carl J. 2005. *India and the Knowledge Economy: Leveraging Strengths and Opportunities.* Washington, DC: World Bank Publications.

Dahlman, Carl J., and Jean Eric Aubert. 2001. *China and the Knowledge Economy: Seizing the 21st Century.* Washington, DC: The World Bank.

Dahlman, Carl, and Anuja Utz. 2005. *India and the Knowledge Economy: Leveraging Strengths and Opportunities.* Washington, DC: The World Bank.

Daneke, Gregory A. 1998. "Beyond Schumpeter: Non-linear Economics and the Evolution of the U.S. Innovation System." *Journal of Socio-economics* 27(1):97-117.

Darby, Michael, Lynne G. Zucker, and Andrew J. Wang. 2002. Program Design and Firm Success in the Advanced Technology Program: Project Structure and Innovation Outcomes. NISTIR 6943. Gaithersburg, MD: National Institute of Standards and Technology.

Das, Gurcharan. 2006. "The India Model." *Foreign Affairs* 85(4).

Dasgupta, P., and P. David. 1994. "Toward a New Economics of Science." *Research Policy* 23:487-521.

David, P. A. 1985. "Clio and the Economics of QWERTY." *American Economic Review* 75(2):332-337.

Davidsson, Per. 1996. "Methodological Concerns in the Estimation of Job Creation in Different Firm Size Classes." Working Paper, Jönköping International Business School.

Davis, Steven, John Haltiwanger, and Scott Schuh. 1993. "Small Business and Job Creation: Dissecting the Myth and Reassessing the Facts." Working Paper No. 4492. Cambridge, MA: National Bureau of Economic Research.

Debackere, Koenraad, and Reinhilde Veugelers. 2005. "The Role of Academic Technology Transfer Organizations in Improving Industry Science Links." *Research Policy* 34(3):321-342.

Debackere, Koenraad, and Wolfgang Glänzel. 2004. "Using a Bibliometric Approach to Support Research Policy Making: The Case of the Flemish BOF-key." *Scientometrics* 59(2).

Defense Advanced Research Projects Agency. 2003. *DARPA Over The Years*. Arlington, VA: Defense Advanced Research Projects Agency. October 27.

Defense Advanced Research Projects Agency. 2005. *DARPA—Bridging the Gap, Powered by Ideas*. Arlington, VA:

Defense Advanced Research Projects Agency. February. de Jonquieres, Guy. 2004. "China and India Cannot Fill the World's Skills Gap." *Financial Times* July 12.

de Jonquieres, Guy. 2004. "To Innovate, China Needs More than Standards." *Financial Times* July 12.

de Jonquieres, Guy. 2006. "China's Curious Marriage of Convenience." *Financial Times* July 19.

De la Mothe, J., and Gilles Paquet. 1998. "National Innovation Systems, 'Real Economies' and Instituted Processes." *Small Business Economics* 11:101-111.

Deng, Xiaoping, General Secretary of the Communist Party of China Central Committee. 1978. Address at the First National Science Congress.

Department of Energy. 2011. *U.S.-China Clean Energy Cooperation: A Progress Report by the U.S. Department of Energy*. January.

De Proft, A. 2006. Presentation at National Academies symposium on "Synergies in Regional and National Policies in the Global Economy." Leuven, Belgium. September.

Dirks, S and M. Keeling. 2009. Executive Report. "A Vision of Smarter Cities: How Cities Can Lead the Way into a Prosperous and Sustainable Future," IBM Global Business Services,

Doloreux, David. 2004. "Regional Innovation Systems in Canada: A Comparative Study." *Regional Studies* 38(5):479-492.

Dries, Ilse, Peer van Humbeek, and Jan Larosse. 2005. *Linking Innovation Policy and Sustainable Development in Flanders*. Paris, France: Organisation for Economic Co-operation and Development.

Dudas, J. 2005. "Statement of the Honorable Jon W. Dudas Deputy Under Secretary of Commerce for Intellectual Property and Director of the U.S. Patent and Trademark Office before the Subcommittee on Intellectual Property, Committee on the Judiciary." U.S. Senate. <*http://judiciary.senate.gov*>.

Eaton, Jonathan, Eva Gutierrez, and Samuel Kortum. 1998. "European Technology Policy." NBER Working Papers 6827. The Economist. 2005. "Competing Through Innovation." December 17.

Edler, J., and S. Kuhlmann. 2005. "Towards One System? The European Research Area Initiative, the Integration of Research Systems and the Changing Leeway of National Policies." Technikfolgenabschätzung: Theorie und Praxis 1(4):59-68.

EE Times. 2006. "Chinese Province Pays to Get 300-mm Wafer Fab." June 28.

Ehlen, Mark A. 1999. *Economic Impacts of Flow-Control Machining Technologies: Early Applications in the Automobile Industry*. NISTIR 6373. Gaithersburg, MD: National Institute of Standards and Technology.

Ehlers, Vernon J. 1998. *Unlocking Our Future: Toward a New National Science Policy, A Report to Congress by the House Committee on Science*. Washington, DC: U.S. Government Printing Office.

Ehrenberg, R. 2000. *Tuition Rising*. Cambridge, MA: Harvard University Press.

Eickelpasch, Alexander, and Michael Fritsch. 2005. "Contests for Cooperation: A New Approach in German Innovation Policy." *Research Policy* 34:1269-1282.

Electronic News. 2006. "SMIC Gets $3B Nod from Chain's Wuhan Government." May 22.

Endquist, Charles, ed. 1997. *Systems of Innovation: Technologies, Institutions, and Organizations*. London, UK: Pinter.

Energy Information Administration. 2009 "State Energy Consumption Estimates: 1960 through 2007," Tables 8-12.

Engardio, Pete. 2009. "Can the Future be Built in America? Inside the U.S. Manufacturing Crisis," *BusinessWeek:* Sept. 21.

EOS Gallup Europe. 2004. Entrepreneurship. Flash Eurobarometer 146. January. Accessed at <http://ec.europa.eu/enterprise/enterprise_policy/survey/eurobaromet er146en.pdf>.

European Commission. 2003. Innovation in Candidate Countries: Strengthening Industrial Performance. Luxembourg: Office for Official Publications of the European Communities, May. European Commission. 2003. "Investing in Research: An Action Plan for Europe 2003." Luxembourg: Office for Official Publications of the European Communities.

European Commission. 2003. *3rd S&T Indicators Report*. Luxembourg: Office for Official Publications of the European Communities.

Evans, D., and Jovanovic, B. 1989. "An Estimated Model of Entrepreneurial Choice under Liquidity Constraints." *Journal of Political Economy* 97:808-827.

Evans, Sir Harold. 2005. *They Made America*. Sloan Foundation Project. New York: Little, Brown and Company.

Executive Office of the President, 2009. "A Strategy for American Innovation: Driving Towards Sustainable Growth and Quality Jobs."<*http://www.whitehouse.gov/assets/documents/SEPT_20__Inn ovation_Whitepaper_FINAL.pdf*>.

Faems, Dries, Bart Van Looy, and Koenraad Debackere. 2005. "Inter-organizational Collaboration and Innovation: Toward a Portfolio Approach." *Journal of Product Innovation Management* 22(3):238-250.

Fan, W., and White, M. J. 2002. "Personal Bankruptcy and the Level of Entrepreneurial Activity." NBER Working Paper, Series 9340. Boston, MA: National Bureau of Economic Research.

Fangerberg, Jan. 2002. *Technology, Growth, and Competitiveness: Selected Essays*. Cheltenham, UK: Edward Elgar.

Federal Communications Commission. 2009. *Connecting America: The National Broadband Plan*. Washington, DC: Federal Communications Commission.

Federal Register Notice. 2004. "2004 WTO Dispute Settlement Proceeding Regarding China: Value Added Tax on Integrated Circuits." April 21.

Federal Trade Commission. 2003. *To Promote Innovation: The Proper Balance of Patent and Competition Law Policy*. Washington, DC: U.S. Government Printing Office. October.

Feigenbaum, Evan. 2003. *China's Techno-Warriors: National Security and Strategic Competition from the Nuclear to the Information Age*. Stanford: Stanford University Press, 2003

Feldman, Maryann and Albert N. Link. 2001. "Innovation Policy in the Knowledge-based Economy." *Economics of Science, Technology and Innovation*. Volume 23. Boston, MA: Kluwer Academic Press.

Feldman, Maryann P., Albert N. Link, and Donald S. Siegel. 2002. *The Economics of Science and Technology: An Overview of Initiatives to Foster Innovation, Entrepreneurship, and Economic Growth*. Boston, MA: Kluwer Academic Press.

Feldman, Maryann, Irwin Feller, Janet Bercovitz, and Richard Burton. 2002. "Equity and the Technology Transfer Strategies of American Research Universities." *Management Science* 48(1):105-121.

Feller, Irwin. 1997. "Technology Transfer from Universities." In John Smart, ed. *Higher Education: Handbook of Theory and Research*. Vol. XII. New York: Agathon Press.

Feller, Irwin. 2004. *A Comparative Analysis of the Processes and Organizational Strategies Engaged in by Research Universities Participating in Industry-University Research Relationships*. Final report submitted to the University of California Industry-University Cooperative Research Program. Agreement No. M-447646-19927-3.

Ferguson, R. and C. Olofsson. 2004. "Science Parks and the Development of NTBFs: Location, Survival and Growth." *Journal of Technology Transfer* 29(1): 5-17.

Field, A. J. 2003. "The Most Technologically Progressive Decade of the Century." *American Economic Review* (September):1406. The Financial Times. 2005. "World Leader in Patents Focuses on Incremental Innovations." October 12.

Flamm, K. 1996. Mismanaged Trade: Strategic Policy and the Semiconductor Industry. Washington, DC: The Brookings Institution.

Flamm, K. 2003. "Microelectronics Innovation: Understanding Moore's Law and Semiconductor Price Trends." *International Journal of Technology, Policy, and Management* 3(2).

Flamm, K. 2003. "The New Economy in Historical Perspective: Evolution of Digital Technology." In *New Economy Handbook*. St. Louis, MO: Academic Press.

Flamm, K. 2003. "SEMATECH Revisited: Assessing Consortium Impacts on Semiconductor Industry R&D." In National Research Council. *Securing the Future: Regional and National Programs to Support the Semiconductor Industry*. Charles W. Wessner, ed. Washington, DC: The National Academies Press.

Florida, Richard, Tim Gulden, and Charlotta Mellander. 2007. "The Rise of the Mega-Region." October.

Foerst, Ann. 2005. *God in the Machine*. New York: Penguin Books.

Fonfria, Antonio, Carlos Diaz de la Guardia, and Isabel Alvarez. 2002. "The Role of Technology and Competitiveness Policies: A Technology Gap Approach." *Journal of Interdisciplinary Economics* 13(1-2-3):223-241.

Fong, Glenn R. 1998. "Follower at the Frontier: International Competition and Japanese Industrial Policy." *International Studies Quarterly.* 42(2).

Fong, Glenn R. 2001. "ARPA Does Windows: The Defense Underpinning of the PC Revolution." *Business and Politics* 3(3).

Foray, Dominique, and Patrick Llerena. 1996. "Information Structure and Coordination in Technology Policy: A Theoretical Model and Two Case Studies." *Journal of Evolutionary Economics* 6(2).

Friedman, Thomas. 2005. *The World Is Flat: A Brief History of the 21st Century.* New York: W. H. Freeman.

Fukugawa, N. 2006. "Science Parks in Japan and Their Value-Added Contributions to New Technology-based Firms." *International Journal of Industrial Organization* 24(2):381-400.

Furman, Jeffrey L., Michael E. Porter, and Scott Stern. 2002. "The Determinants of National Innovative Capacity." *Research Policy* 31:899-933.

Gallaher, M. P., A. N. Link, and J. E. Petrusa. 2006. *Innovation in the U.S. Service Sector.* London, UK: Routledge.

Geiger, R. 1986. *To Advance Knowledge.* New York: Oxford University Press.

Geiger, R. 1993. *Research and Relevant Knowledge.* New York: Oxford University Press.

Geithner, Timothy. 2010. Joint Press Availability in Beijing, China. May 25.

George, Gerard, and Ganesh N. Prabhu. 2003. "Developmental Financial Institutions as Technology Policy Instruments: Implications for Innovation and Entrepreneurship in Emerging Economies." *Research Policy* 32(1):89-108.

Gibb, M. J. 1985. *Science Parks and Innovation Centres: Their Economic and Social Impact.* Amsterdam, The Netherlands: Elsevier.

Goldstein, H. A., and M. I. Luger. 1990. "Science/Technology Parks and Regional Development Theory." *Economic Development Quarterly* 4(1):64-78.

Goldstein, H. A., and M. I. Luger. 1992. "University-based Research Parks as a Rural Development Strategy." *Policy Studies Journal* 20(2):249-263.

Goodwin, James C., et al. 1999. *Technology Transition*. Arlington, VA: Defense Advanced Research Projects Agency.

Government of the People's Republic of China. 2004. *Anticompetitive Practices of Multinational Companies and Countermeasures*. Administration of Industry and Commerce, Office of Antimonopoly, Fair Trade Bureau, State Administration of Industry and Commerce. May.

Government of the People's Republic of China. 2005. *Article 10: Exemptions of Monopoly Agreements*. Anti-Monopoly Law of the People's Republic of China. Revised July 27.

Government of the People's Republic of China, Ministry of Information Industry. 2006. "Outline of the 11th Five-Year Plan and Medium-and-Long-Term Plan for 2020 for Science and Technology Development in the Information Industry." Xin Bu Ke, No. 309, August 29.

Government of the People's Republic of China, National Development and Reform Commission. 2006. *The 11th Five-Year Plan*. March 19. <http://english.gov.cn/2006-07/26/content_346731.htm>.

Grande, Edgar. 2001. "The Erosion of State Capacity and European Innovation Policy: A Comparison of German and EU Information Technology Policies." *Research Policy* 30(6):905-921.

Grayson, L. 1993. *Science Parks: An Experiment in High-Technology Transfer*. London, UK: The British Library Board.

Griffing, Bruce. 2001. "Between Invention and Innovation, Mapping the Funding for Early-Stage Technologies." Presentation at Carnegie Conference Center. Washington, DC. January 25.

Griliches, Z. 1993. "The Search for R&D Spillovers." *Scandinavian Journal of Economics* 94(S): S29-S47.

Grindley, Peter, David Mowery, and Brian Silverman. 1994. "SEMATECH and Collaborative Research: Lessons in the Design of High Technology Consortia." *Journal of Policy Analysis and Management* 13(4):723-758.

Grove, A. 2001. Swimming Across. New York: Warner Books.

Gu, Shulin and Lundvall Bengt-Åke. 2006. "Policy learning as a key process in the transformation of the Chinese Innovation Systems." In *Asian Innovation Systems in Transition*. Edward Elgar Publishing Ltd.

Guo Ban Han No. 30. 2006. Letter from the General Office of the State Council on Approving the Formulation of the Rules for Implementation of the Several Supporting Policies for Implementation of the Outline of the National Medium and Long-term Plan for Development of Science and Technology. *Gazette of the State Council.* Issue No. 17, Serial No. 1196, June 20.

Guy, I. 1996. "A Look at Aston Science Park." *Technovation* 16(5):217-218.

Guy, I. 1996. "New Ventures on an Ancient Campus." *Technovation* 16(6):269-270.

Hackett, S. M., and D. M. Dilts. 2004. "A Systematic Review of Business Incubation Research." *Journal of Technology Transfer* 29(1):55-82.

Hall, B. H. 2002. "The Assessment: Technology Policy." *Oxford Review of Economic Policy* 18(1):1-9.

Hall, B. H. 2005. "Exploring the Patent Explosion." *Journal of Technology Transfer* 30:35-48.

Hall, B. H., A. N. Link, and J. T. Scott. 2001. "Barriers Inhibiting Industry from Partnering with Uni-versities: Evidence from the Advanced Technology Program." *Journal of Technology Transfer* 26(1-2):87-98.

Hall, B. H., A. N. Link, and J. T. Scott. 2003. "Universities as Research Partners." *Review of Economics and Statistics* 85(2):485-491.

Hall, B. H., A. Jaffe, and M. Trajtenberg. 2005. "Market Value and Patent Citations." *RAND Journal of Economics* 36:16-38.

Hall, B. H., and R. H. Ziedonis. 2001. "The Patent Paradox Revisited: An Empirical Study of Patenting in the U.S. Semiconductor Industry. 1979-1995." *RAND Journal of Economics* 32:101-128.

Halpern, Mark. 2006. "The Trouble with the Turing Test." *The New Atlantis* 11(Winter):42-63.

Hane, G. 1999. "Comparing University-Industry Linkages in the United States and Japan." In L. Branscomb, F. Kodama, and R. Florida, eds. *Industrializing Knowledge.* Cambridge, MA: The MIT Press.

Hansson, F., K. Husted, and J. Vestergaard. 2005. "Second Generation Science Parks: From Structural Holes Jockeys to Social Capital Catalysts of the Knowledge Society." *Technovation* 25(9):1039-1049.

Hart, David. 1998. *Forged Consensus.* Princeton, NJ: Princeton University Press.

Hashimoto, T. 1999. "The Hesitant Relationship Reconsidered: University-Industry Cooperation in Postwar Japan." In Lewis M. Branscomb, Fumio Kodama, and Richard Florida, eds. *Industrializing Knowledge*: University-Industry Linkage in Japan and the United States. Cambridge, MA: The MIT Press.

Hassan, Mohamed H.A. 2005. Small Things and Big Changes in the Developing World." *Science* 309: 65-66 (July 1)

Hayashi, Fumio, and Edward C. Prescott. 2002. "The 1990s in Japan: A Lost Decade." *Review of Economic Dynamics* 5(1):206-235.

Heller, M., and R. Eisenberg. 1998. "Can Patents Deter Innovation? The Anticommons in Biomedical Research." *Science* 280:698.

Henderson, J. V. 1986. "The Efficiency of Resource Usage and City Size." *Journal of Urban Economics* 19(1):47-70.

Henderson, Jennifer A., and John J. Smith. 2002. "Academia, Industry, and the Bayh-Dole Act: An Implied Duty to Commercialize." White Paper. Center for the Integration of Medicine and Innovative Technology. Harvard University. October.

Henderson, R., L. Orsenigo, and G. P. Pisano. 1999. "The Pharmaceutical Industry and the Revolution in Molecular Biology: Interactions among Scientific, Institutional, and Organizational Change." In D. C. Mowery and R. R. Nelson, eds. *Sources of Industrial Leadership*. Cambridge, UK: Cambridge University Press.

Henderson, R., M. Trajtenberg, and A. Jaffe. 1998. "Universities as a Source of Commercial Technology: A Detailed Analysis of University Patenting, 1965-1988." *Review of Economics and Statistics* 80(1):119-127.

Hennessy, J. L., and D. A. Patterson 2002. *Computer Architecture: A Quantitative Approach*. 3rd edition. San Francisco, CA: Morgan Kaufmann Publishers Inc.

Hicks, D., T. Breitzman, D. Olivastro, and K. Hamilton. 2001. "The Changing Composition of Innovative Activity in the U.S.—A Portrait Based on Patent Analysis." *Research Policy* 30(4):681-704.

Hilpert, U., and B. Ruffieux. 1991. "Innovation, Politics and Regional Development: Technology Parks and Regional Participation in High-Technology in France and West Germany." In U. Hilpert, ed. *Regional Innovation and Decentralization: High-Technology Industry and Government Policy*. London, UK: Routledge.

Himmelberg, C., and B. C. Petersen. 1994. "R&D and Internal Finance: A Panel Study of Small Firms in High-Tech Industries." *Review of Economics and Statistics* 76(1):38-51.

Holtz-Eakin, D., D. Joulfian, and H. Rosen. 1994. "Striking it Out: Entrepreneurial Survival and Liquidity Constraints." *Journal of Political Economy* 102(1):53-75.

Hoshi, Takeo, and Anil Kashyap. 2004. "Japan's Economic and Financial Crisis: An Overview." *The Journal of Economic Perspectives* Winter.

Howell, Thomas. 2003. "Competing Programs: Government Support for Microelectronics." In National Research Council. *Securing the Future: Regional and National Programs to Support the Semiconductor Industry.* Charles W. Wessner, ed. Washington, DC: The National Academies Press.

Howell, Thomas, et al. 2003. *China's Emerging Semiconductor Industry.* San Jose, CA: Semiconductor Industry Association, October.

Hu, Jintao. General Secretary of the Communist Party of China Central Committee. 2005. Keynote Speech, November 27.

Hu, Jintao. General Secretary of the Communist Party of China Central Committee. 2007. Speech to the Seventeenth Communist Party of China National Conference, October.

Hu, Jintao. President of China. 2009. Speech to United Nations General Assembly, September 22.

Hu, Zhijian. 2006. "IPR Policies In China: Challenges and Directions." Presentation at Industrial Innovation in China. Levin Institute Conference. July 24-26.

Huang, Can, C. Amorim, M. Spinoglio, B. Gouveia and A. Medina. 2004. "Organization, Programme and Structure:An Analysis of the Chinese Innovation Policy Framework." *R&D Management* 34(4).

Huang, Yasheng, and Tarun Khanna. 2003."Can India Overtake China?" *Foreign Policy.* July-August.

Huddleson, Lillian, and Vicki Daitch. 2002. *True Genius—The Life and Science of John Bardeen.* Washington, DC: Joseph Henry Press.

Hughes, Kent. 2005. *Building the Next American Century: The Past and Future of American Economic Competitiveness.* Washington, DC: Woodrow Wilson Center Press. Chapter 14.

Hughes, Kent H. 2005. "Facing the Global Competitiveness Challenge." *Issues in Science and Technology* XXI(4):72-78.

Hulsink, W., H. Bouwman, and T. Elfring. 2007. "Silicon Valley in the Polder? Entrepreneurial Dynamics, Virtuous Clusters and Vicious Firms in the Netherlands and Flanders." ERIM Report Series Research in Management (ERS-2007-048-ORG).

Hundley, Richard O. 1999. *Past Revolutions, Future Transformations: What Can the History of Revolutions in Military Affairs Tell Us About Transforming the U.S. Military.* Rand Corporation, National Research Institute.

Iansati, Marco, and Roy Levien. 2005. *The Keystone Advantage.* Cambridge, MA: Harvard Business School Press.

Industrial Research Institute, Inc. 2001. "Industry-University Intellectual Property." Position Paper. External Research Directors Network. April.

Inkpen, Andrew C. and Wang Pien. 2006. "An Examination of Collaboration and Knowledge Transfer: China-Singapore Suzhou Industrial Park." *Journal of Management Studies* 43(4):779-811.

Inside U.S.-China Trade. 2006. "Industry Worried China Backing out of Commitment to Join GPA." September 27.

Institute for International Education. 2011. "International Student Enrollments Rose Modestly in 2009/10, Led by Strong Increase in Students from China." Press Release *<http://www.iie.org/en>*.

International Association of Science Parks (IASP). 2000. *<http://www.iaspworld.org/information/definitions.php>*.

International Telecommunications Union. 2005. "ITU Internet Reports 2005: The Internet of Things," *<http://www.itu.int/osg/spu/publications/internetofthings/InternetofThings_summary.pdf>*.

Jacobs, Tom. 2002. "Biotech Follows Dot.com Boom and Bust." *Nature* 20(10):973.

Jaffe, A. B. 1989. "Real Effects of Academic Research." *American Economic Review* 79(5): 957-970.

Jaffe, A. B. 1997. *Economic Analysis of Research Spillovers: Implications for the Advanced Technology Program.* NIST GCR 97-708. Gaithersburg, MD: National Institute of Standards and Technology.

Jaffe, A. B. 1998. "The Importance of 'Spillovers' in the Policy Mission of the ATP." *Journal of Technology Transfer* 23(1):11-19.

Jaffe, A. B. 2000. "The U.S. Patent System in Transition: Policy Innovation and the Innovation Process." *Research Policy* 29:531-557.

Jaffe, A. B., and J. Lerner. 2006. *Innovation and Its Discontents: How Our Broken Patent System Is Endangering Innovation and Progress, and What to Do About It.* Princeton, NJ: Princeton University Press.

Jaffe, A. B., J. Lerner, and S. Stern, eds. 2003. *Innovation Policy and the Economy.* Volume 3. Cambridge, MA: The MIT Press.

Jaffe, A. B., M. Trajtenberg, and R. Henderson. 1993. "Geographic Localization of Knowledge Spillovers as Evidenced by Patent Citations." *Quarterly Journal of Economics* 108(3):577-598.

Japan Patent Office. 2003. *Reports on Technology Trend and Patent Application: Life Science.* (in Japanese).

Jasanoff, Sheila, ed. 1997. Comparative Science and Technology Policy. Elgar Reference Collection. *International Library of Comparative Pubic Policy.* Volume 5. Cheltenham, UK: Edward Elgar.

Jorgenson, D. W. 2001. "Information Technology and the U.S. Economy." *American Economic Review* 91(1) March.

Jorgenson, D. W. 2001. "U.S. Economic Growth in the Information Age." *Issues in Science and Technology* Fall.

Jorgenson, D.W., and Kevin Stiroh. 2002. "Raising the Speed Limit: Economic Growth in the Information Age." In National Research Council. *Measuring and Sustaining the New Economy.* Dale Jorgenson and Charles Wessner, eds. Washington, DC: The National Academies Press.

Jorgenson, D. W., and K. Motohashi. 2003. "The Role of Information Technology in the Economy: Comparison between Japan and the United States." Prepared for RIETI/KEIO Conference on Japanese Economy: Leading East Asia in the 21st Century? Keio University. May 30.

Joy, William. 2000. "Why the Future Does Not Need Us." *Wired* 8.04. April.

Kaelble, Steve. 2004. "Good Neighbors: Indiana's Certified-Technology-park Program." *Indiana Business Magazine* July 1.

Kapur, Devesh. 2003. "Indian Diaspora as a Strategic Asset." *Economic and Political Weekly* 38(5):445-448. The Kauffman Foundation and the Information Technology and Innovation Foundation. 2007. *The 2007 State New Economy Index.* <*http://www.kauffman.org/items.cfm?itemID=766*>.

Kawamoto, Takuji. 2004. "What Do the Purified Solow Residuals Tell Us about Japan's Lost Decade?" Bank of Japan IMES Discussion Paper Series. No. 2004-E-5. Tokyo, Japan: Bank of Japan.

Kenney, Martin, ed. 2000. Understanding Silicon Valley: The Anatomy of an Entrepreneurial Region. Stanford, CA: Stanford University Press.

Kim, Linsu. 1997. *Imitation to Innovation; The Dynamics of Korea's Technological Learning.* Boston: Harvard Business School Press. pp. 192-213, 234-243.

Kim, Yong-June. 2006. "A Korean Perspective on China's Innovation System." Presentation at Industrial Innovation in China. Levin Institute Conference. July 24-26.

Kneller, R. 2003. "University-Industry Cooperation and Technology Transfer in Japan Compared with the U.S.: Another Reason for Japan's Economic Malaise?" *Journal of International Economic Law* 24:329-450.

Koizumi, Kei. 2007. "Historical Trends in Federal R&D." In AAAS Report XXXII: Research and Development FY2008. AAAS Publication Number 07-1A. Washington, DC: American Association for the Advancement of Science.

Kondo, Masayuki. 2003. "Chinese Model to Create High-Tech Start-Ups from Universities and Research Institutes." In M. von Zedtwitz, G. Haour, T. Khalil, and L. Lefebvre, eds. *Management of Technology: Growth through Business, Innovation and Entrepreneurship.* Oxford, UK: Pergamon Press.

Kondo, Masayuki. 2004. "Policy Innovation in Science and Technology in Japan—from S&T Policy to Innovation Policy." (In Japanese.) *Journal of Science Policy and Research Management* 19(3/4):132-140.

Kondo, Masayuki. 2004. "University spin-offs in Japan." *Asia Pacific Tech Monitor.* March-April 2004. Pp. 37-43. Asian and Pacific Centre for Transfer of Technology, ESCAP, UN.

Kondo, Masayuki. 2005. "Spin-offs from Public Research Institutes as Domestic Technology Transfer Means—The Case of RIKEN and Riken Industrial Group." (In Japanese.) *Development Engineering* 11:31-41.

Kondo, Masayuki. 2006. "University-Industry Partnerships in Japan." Presentation at the conference, 21st Century Innovation Systems for the United States and Japan: Lessons from a Decade of Change. Tokyo, Japan. The National Academies, NISTEP of Japan, and The Institute of Innovation Research of Hitotsubashi University. *<http://www.nistep.go.jp/IC/ic060110/pdf/5-2.pdf>.*

Kortum, S., and J. Lerner. 1999. "What is Behind the Recent Surge in Patenting?" *Research Policy* 28:1-22.

Koschatzky, Knut. 2003. "The Regionalization of Innovation Policy: New Options for Regional Change?" In G. Fuchs and Phil Shapira, eds. *Rethinking Regional Innovation: Path Dependency or Regional Breakthrough?* London, UK: Kluwer.

Krugman, Paul. 1991. *Geography and Trade.* Cambridge, MA: MIT Press.

Krugman, Paul. 1994. "Competitiveness: A Dangerous Obsession." *Foreign Affairs* March/April.

Kuhlmann, Stephan and Jakob Edler. 2003. "Scenarios of Technology and Innovation Policies in Europe: Investigating Future Governance—Group of 3." *Technological Forecasting & Social Change* 70.

Kumar, Deepak. 1995. *Science and the Raj: 1857-1905.* New York: Oxford University Press.

Lall, Sanjaya. 2002. "Linking FDI and Technology Development for Capacity Building and Strategic Competitiveness." *Transnational Corporations* 11(3):39-88.

Lanjouw, J. O., and I. Cockburn. 2000. "Do Patents Matter? Empirical Evidence after GATT." NBER Working Paper No. 7495.

Lanjouw, J. O., and J. Lerner 1997. "The Enforcement of Intellectual Property Rights: A Survey of the Empirical Literature." NBER Working Paper No. W6296. Available at <http://ssrn.com/abstract=226053>.

Laredo, Philippe, and Philippe Mustar, eds. 2001. *Research and Innovation Policies in the New Global Economy: An International Perspective.* Cheltenham, UK: Edward Elgar.

Larosse, J. 2004. *Towards a 'Third Generation' Innovation Policy in Flanders: Policy Profile of the Flemish Innovation.* Brussels, Belgium: IWT.

Lazowska, Edward D., and David Paterson. 2005. "An Endless Frontier Postponed." *Science* 308(5723):757. May 6.

Lebra, Takie Sugiyama. 1971. "The Social Mechanism of Guilt and Shame: The Japanese Case." *Anthropological Quarterly* 44(4):241-255.

Lembke, Johan. 2002. *Competition for Technological Leadership: EU Policy for High Technology.* Cheltenham, UK: Edward Elgar.

Lemola, Tarmo. 2002. "Convergence of National Science and Technology Policies: The Case of Finland." *Research Policy* 31(8-9):1481-1490.

Lerner, J. 1995. "Patenting in the Shadow of Competitors." *Journal of Law and Economics* 38(Oct):463-495.

Lerner, J. 1999. "Public Venture Capital." In National Research Council. *The Small Business Innovation Program: Challenges and Opportunities.* Charles W. Wessner, ed. Washington, DC: National Academy Press.

Lerner, J. 2002. "Patent Policy Shifts and Innovation Over 150 Years." *American Economic Review P&P.* (May).

Lerner, J., and J. Tirole. 2004. "Efficiency of Patent Pools." *American Economic Review* 94(3):691-711.

Leslie, Stuart, and Robert Kargon. 2006. "Exporting MIT." *Osiris* 21:110-130.

Levin, R. C., A. K. Klevorick, R. R. Nelson, and S. G. Winter. 1987. "Appropriating the Returns to Industrial R&D." *Brookings Papers on Economic Activity* 783-820.

Levitt, Rachelle, ed. 1987. *The University/Real Estate Connection: Research Parks and Other Ventures.* Washington, DC: Urban Land Institute.

Lewis, James A. 2005. *Waiting for Sputnik: Basic Research and Strategic Competition.* Washington, DC: Center for Strategic and International Studies.

Leyden, D. P., A. N. Link, and D. S. Siegel. 2008. "A Theoretical and Empirical Analysis of the Decision to Locate on a University Research Park." *IEEE Transactions on Engineering Management* 55(1):23-28.

Li, Jianjun. 2006. "The Development and Opening of Tianjin Binhai: New Area & China's Biotechnical Innovations." Presentation at Industrial Innovation in China. Levin Institute Conference. July 24-26.

Licklider, J. C. R. 1960. "Man-Computer Symbiosis." *IRE Transactions on Human Factors in Electronics.* March.

Lin, Otto. 1998. "Science and Technology Policy and its Influence on the Economic Development of Taiwan." In Henry S. Rowen, ed. Behind *East Asian Growth: The Political and Social Foundations of Prosperity.* New York: Routledge.

Lindelöf, P., and H. Löfsten. 2003. "Science Park Location and New Technology-Based Firms in Sweden: Implications for Strategy and Performance." *Small Business Economics* 20(3):245-258.

Lindelöf, P. and H. Löfsten. 2004. "Proximity as a Resource Base for Competitive Advantage: University-Industry Links for Technology Transfer." *Journal of Technology Transfer* 29(3-4): 311-326.

Lindh, T., and Ohlsson, H. 1996. "Self-Employment and Windfall Gains: Evidence from the Swedish Lottery." *Economic Journal* 106:1515-1526.

Link, A. N. 1981. "Basic Research and Productivity Increase in Manufacturing: Some Additional Evidence." *American Economic Review* 71(5):1111-1112.

Link, A. N. 1981. *Research and Development Activity in U.S. Manufacturing.* New York: Praeger.

Link, A. N. 1995. *A Generosity of Spirit: The Early History of the Research Triangle Park.* Research Triangle Park, NC: University of North Carolina Press for the Research Triangle Park Foundation.

Link, A. N. 2002. *From Seed to Harvest: The Growth of the Research Triangle Park.* Research Triangle Park, NC: University of North Carolina Press for the Research Triangle Park Foundation.

Link, A. N., and D. S. Siegel. 2003. *Technological Change and Economic Performance.* London, UK: Routledge.

Link, A. N., and J. T. Scott. 1998. *Public Accountability: Evaluating Technology-Based Institutions.* Norwell, MA: Kluwer Academic Publishers.

Link, A. N., and J. T. Scott. 2001. "Public/Private Partnerships: Stimulating Competition in a Dynamic Market." *International Journal of Industrial Organization* 19(5):763-794.

Link, A. N., and J. T. Scott. 2003. "The Growth of Research Triangle Park." *Small Business Economics* 20(2):167-175.

Link, A. N., and J. T. Scott. 2003. "U.S. Science Parks: The Diffusion of an Innovation and Its Effects on the Academic Mission of Universities." *International Journal of Industrial Organization* 21(9):1323-1356.

Link, A. N., and J. T. Scott. 2005. "Opening the Ivory Tower's Door: An Analysis of the Determinants of the Formation of U.S. University Spin-Off Companies." *Research Policy* 34(7):1106-1112.

Link, A. N., and J. T. Scott. 2006. "U.S. University Research Parks." *Journal of Productivity Analysis* 25(1):43-55.

Link, A. N., and J. T. Scott. 2007. "The Economics of University Research Parks." *Oxford Review of Economic Policy* 23(4):661-674.

Link, A. N., and K. R. Link. 2003. "On the Growth of U.S. Science Parks." *Journal of Technology Transfer* 28(1):81-85.

Litan, Robert E., and Lesa Mitchell. 2008. "Should Universities be Agents of Economic Development?" *Astra Briefs* 7(7-8).

Locke, Gary. 2010. "Remarks on Clean Energy Trade Mission to China, Indonesia." Briefing at the Washington Foreign Press Center: May 12.

Löfsten, Hans, and Peter Lindelöf. 2002. "Science Parks and the Growth of New Technology-based Firms—Academic-industry Links, Innovation and Markets." *Research Policy* 31(6):859-876.

Long, Guogiang. 2005. "China's Policies on FDI: Review and Evaluation." In Theodore H. Moran, Edward M. Graham, and Magnus Blomström, eds. *Does Foreign Direct Investment Promote Development?* Washington, DC: Institute for International Economics.

Luger, M. I. 2001. "Introduction: Information Technology and Regional Economic Development." *Journal of Comparative Policy Analysis: Research & Practice.*

Luger, M. I. 2007. "Smart Places for Smart People: Using Cluster-based Planning in the 21st Century." In *Creating Enterprise: Igniting Innovation through Business-University-Government Networks.* Cheltenham, UK: Edward Elgar.

Luger, M. I., and H. A. Goldstein. 1991. *Technology in the Garden.* Chapel Hill, NC: University of North Carolina Press.

Luger, M. I., and H. A. Goldstein. 2006. *Research Parks Redux: The Changing Landscape of the Garden.* Washington, DC: U.S. Economic Development Administration.

Lundstörm, A., and L. Stevenson. 2001. *Entrepreneurship Policy for the Future.* Stockholm, Sweden: FSF.

Lynn, Leonard. 2006. "Collaborative Advantage and China's Evolving Position in the Global Technology System." Presentation at Industrial Innovation in China. Levin Institute Conference. July 24-26.

Macher, Jeffrey, David Mowery, and David Hodges. 1999. "Semiconductors." In National Research Council. *U.S. Industry in 2000: Studies in Competitive Performance.* David C. Mowery, ed. Washington, DC: National Academy Press.

Maddison, Angus, and Donald Johnston. 2001. *The World Economy: A Millennial Perspective.* Paris, France: Organization for Economic Co-operation and Development.

Malecki, E. J. 1991. *Technology and Economic Development.* New York: John Wiley.

Mani, Sunil. 2004. "Government, Innovation and Technology Policy: An International Comparative Analysis." *International Journal of Technology and Globalization* 1(1).

Mansfield, E. 1986. "Patents and Innovation: An Empirical Study." *Management Science* 32:173-181.

Mansfield, E. 1996. *Estimating Social and Private Returns from Innovations Based on the Advanced Technology Program: Problems and Opportunities.* GCR 99-780. Gaithersburg, MD: National Institute of Standards and Technology.

Mansfield, E. 1996. "How Fast Does New Industrial Technology Leak Out?" *Journal of Industrial Economics* 34(2):217-224.

Marshall, Andrew. 1993. "Some Thoughts on Military Revolutions-Second Version." Memorandum for the Record. August 23.

Mazuzan, George. 1988. *The National Science Foundation: A Brief History (1950-1985)*. NSF 88-16. Arlington, VA: The National Science Foundation.

McKibben, William. 2003. *Enough: Staying Human in an Engineered Age*. New York: Henry Holt & Co.

Megginson, William L. 2004. "Towards a Global Model of Venture Capital?" *Journal of Applied and Corporate Finance* 16(1).

Merges, R. P. 1996. "A Comparative Look at Intellectual Property Rights and the Software Industry." In David C.

Mowery, ed. *The International Software Industry*. New York: Oxford University Press.

Merges, R. P. 1999. "As Many as Six Impossible Patents before Breakfast: Property Rights for Business Concepts and Patent System Reform." Berkeley Technology Law Journal.

Merges, R. P., and R. R. Nelson. 1990. "On the Complex Economics of Patent Scope." *Columbia Law Review* 90:839-916.

Merton, R. K. 1968. "The Matthew Effect in Science." *Science* 159(3810):56-63.

METI (Small and Medium Enterprise Agency, Ministry of Economy, Trade and Industry). 2003. *The 2003 White Paper on Small and Medium Enterprises in Japan*. Tokyo: Japan Small Business Research Institute.

METI. 2005. *The 2005 White Paper on Small and Medium Enterprises in Japan*. Tokyo: Japan Small Business Research Institute.

MEXT. 2004. *Annual Report on Science and Technology Promotion Measures 2004: White Paper on Science and Technology*. (Heisei 16 nendo Kagaku-Gijutu Sinkou-ni-kansuru Nenji-houkoku). MEXT.

Meyer-Krahmer, Frieder. 2001. "The German Innovation System." Pp. 205-252 in P. Larédo and P. Mustar, eds. *Research and Innovation Policies in the New Global Economy: An International Comparative Analysis*. Cheltenham, UK: Edward Elgar.

Meyer-Krahmer, Frieder. 2001. "Industrial Innovation and Sustainability—Conflicts and Coherence." Pp. 177-195 in Daniele Archibugi and Bengt-Ake Lundvall, eds. *The Globalizing Learning Economy*. New York: Oxford University Press.

Middleton, Andrew, and Steven Bowns. With Keith Hartley and James Reid. 2006. "The Effect of Defense R&D on

Military Equipment Quality." *Defense and Peace Economics* 17(2):117-139. April.

Miller, Roger, and Marcel Cote. 1987. *Growing the Next Silicon Valley: A Guide for Successful Regional Planning.* Toronto: D. C. Heath and Company.

Mills, K, E. Reynolds, and A Reamer. 2008. *Clusters and Competitiveness: A New Federal Role for Stimulating Regional Economies.* Washington, DC: Brookings Institution.

Ministry of Economy, Trade and Industry. 2006. *New Economic Growth Strategies 2006.* Tokyo, Japan: Research Institute of Economy, Trade and Industry. (In Japanese).

Ministry of Health, Labor and Welfare. 2002. *Visions of Pharmaceutical Industry in Japan.* *<http://www.mhlw.go.jp/houdou/2002/04/h0409-1.html>* (In Japanese).

Mitra, Raja. 2006. "India's Potential as a Global R&D Power." In Magnus Karlsson, ed. *The Internationalization of Corporate R&D.* Östersund: Swedish Institute for Growth Policy Studies.

Miyata, Shinpei. 1983. *A Free Paradise for Scientists.* (In Japanese.) Bungeishunju.

Mody, Ashok, and Carl Dahlman. 1992. "Performance and Potential of Information Technology: An International Perspective." *World Development* 20(12):1703-1719.

Moore, Gordon. 2003. "The SEMATECH Contribution." In National Research Council. *Securing the Future: Regional and National Programs to Support the Semiconductor Industry.* Charles W. Wessner, ed. Washington, DC: The National Academies Press.

Mooris, P. R. 1990. *A History of The World Semiconductor Industry.* Stevenage, UK: Peter Peregrinus Ltd.

Morrison, Wayne M. 2010. "China-U.S. Trade Issues" Washington, DC: Congressional Research Service: June 1.

Morrow, Daniel S. 2003. Dr. J. Craig Venter-Oral History. Computer World Honors Program. April 21. Available at <http://cwheroes.org/archives/histories/venter>.

Moser, P. 2001. "How Do Patent Laws Influence Innovation? Evidence from Nineteenth Century World Fairs." University of California at Berkeley: working paper.

Motohashi, Kazuyuki. 2005. "University-industry Collaborations in Japan: The Role of New Technology-based Firms in Transforming the National Innovation System." *Research Policy* 34:583-594.

Motohashi, Kazuyuki, and Xiao Yun. 2007. "China's Innovation System Reform and Growing Industry and Science Linkages." *Research Policy* 36:1251-1260.

Mowery, D. C., and A. A. Ziedonis. 2002. "Academic Patent Quality and Quantity Before and After the Bayh-Dole Act in the United States." *Research Policy* 31(3):399-418.

Mowery, D. C., and B. N. Sampat. 2005. "The Bayh-Dole Act of 1980 and University-Industry Technology Transfer: A Model for Other OECD Governments?" *Journal of Technology Transfer* 30(1/2):115-127.

Mowery, D. C., R. Nelson, B. Sampat, and A. Ziedonis. 2001. "The Growth of Patenting and Licensing by U.S.

Universities: An Assessment of the Effects of the Bayh-Dole Act of 1980." *Research Policy* 30.

Mowery, D. C., R. Nelson, B. Sampat, and A. Ziedonis. 2003. The Ivory Tower and Industrial Innovation. Stanford, CA: Stanford University Press.

Muro, Mark and S. Rahman. 2010. "Budget 2011: Industry Clusters as a Paradigm for Job Growth," Brookings Institution Metropolitan Policy Program. June 10.

Murphy, L. M., and P. L. Edwards. 2003. Bridging the Valley of Death: Transitioning from Public to Private Sector Financing. Golden, CO: National Renewable Energy Laboratory. May.

Mustar, Phillipe, and Phillipe Laredo. 2002. "Innovation and Research Policy in France (1980-2000) or the Disappearance of the Colbertist State." *Research Policy* 31:55-72.

Nadiri I. 1993. *Innovations and Technological Spillovers*. NBER Working Paper No. 4423. Boston, MA: National Bureau of Economic Research.

Nagaoka, Sadao. 2005. "How Does Priority Rule Work? Evidence from the Patent Examination Records in Japan." A Paper Presented for Patent Statistics and Innovation Research Workshop, November 25. Research Center for Advanced Science and Technology, University of Tokyo.

Nagaoka, Sadao. 2006. "Reform of Patent System in Japan and Challenges." Presentation at the conference, "21st Century Innovation Systems for the United States and Japan: Lessons from a Decade of Change." Tokyo, Japan. The National Academies, NISTEP of Japan, and The Institute of Innovation Research of Hitotsubashi University. *<http://www.nistep.go.jp/IC/ic060110/pdf/5-2.pdf>*.

Nagaoka, Sadao. 2007. "Assessing the R&D Management of Firms by Patent Citation: Evidence from the U.S. Patents." Journal of Economics & Management Strategy 16(1).

Nagaoka, Sadao, Naotoshi Tsukada, and Tomoyuki Shimbo. 2006. "The Emergence and Structure of Essential Patents of Standards: Lessons from Three IT Standards." IIR Working Paper WP#06-08. Institute of Innovation Research, Hitotsubashi University.

Nakamura, K., Y. Okada, and A. Tohei. 2006. "Does the Public Sector Make a Significant Contribution to Biomedical Research in Japan? A Detailed Analysis of Government and University Patenting, 1991-2002." Discussion Paper Series. Competition Policy Research Center. Fair Trade Commission of Japan. CPDP25-E. <http://www.jftc.go.jp/cprc/DP/discussionpapers.html>.

Nakayama, Yasuo, Mitsuaki Hosono, Nobuya Fukugawa, and Masayuki Kondo. 2005. University-Industry Cooperation: Joint Research and Contract Research. (In Japanese.) NISTEP Research Material. No. 119. Tokyo, Japan: National Institute of Science and Technology Policy, Ministry of Education, Culture, Sports, Science and Technology.

NASSCOM Research. 2007. White Paper. "Tracing China's IT Software and Services Industry Evolution".

National Academy of Engineering. 2004. The Engineer of 2020: Visions of Engineering in the New Century. Washington, DC: The National Academies Press.

National Academy of Sciences/National Academy of Engineering/Insitute of Medicine. 2007. Rising Above the Gathering Storm: Energizing and Employing America for a Brighter Economic Future. Washington, DC: The National Academies Press.

National Institute of Science and Technology Policy and Mitsubishi Research Institute. 2005. Government S&T Budget Analysis during the First and Second S&T Basic Plans. (Dai-ikki oyobi dai-niki Kagaku-Gijutu-Kihonkeikaku kikanchuu no kenkyuu-kaihatu-soushi no naiyou-bunseki). NISTEP Report No. 84. March.

National Manufacturing Competitiveness Council. 2006. "The National Strategy for Manufacturing." New Delhi, India. March.

National Research Council. 1982. Scientific Communication and National Security. Washington, DC: National Academy Press.

National Research Council. 1987. Balancing the National Interest: U.S. National Security Export Controls and Global Economic Competition. Washington, DC: National Academy Press.

National Research Council. 1996. *Conflict and Cooperation in National Competition for High-technology Industry.* Washington, DC: National Academy Press.

National Research Council. 1999. *The Advanced Technology Program: Challenges and Opportunities.* Charles W. Wessner, ed. Washington, DC: National Academy Press.

National Research Council. 1999. *Funding a Revolution: Government Support for Computing Research.* Washington, DC: National Academy Press.

National Research Council. 1999. *Industry-Laboratory Partnerships: A Review of the Sandia Science and Technology Park Initiative.* Charles W. Wessner, ed. Washington, DC: National Academy Press.

National Research Council. 1999. *New Vistas in Transatlantic Science and Technology Cooperation.* Charles W. Wessner, ed. Washington, DC: National Academy Press.

National Research Council. 1999. *The Small Business Innovation Research Program: Challenges and Opportunities.* Charles W. Wessner, ed. Washington, DC: National Academy Press.

National Research Council. 2000. *The Small Business Innovation Research Program: A Review of the Department of Defense Fast Track Initiative.* Charles W. Wessner, ed. Washington, DC: National Academy Press.

National Research Council. 2000. *U.S. Industry in 2000: Studies in Competitive Performance.* David C. Mowery, ed. Washington, DC: National Academy Press.

National Research Council. 2001. *The Advanced Technology Program: Assessing Outcomes.* Charles W. Wessner, ed. Washington, DC: National Academy Press.

National Research Council. 2001. *Building a Workforce for the Information Economy.* Washington, DC: National Academy Press.

National Research Council. 2001. *Capitalizing on New Needs and New Opportunities: Government-Industry Partnerships in Biotechnology and Information Technologies.* Charles W. Wessner, ed. Washington, DC: National Academy Press.

National Research Council. 2001. *A Review of the New Initiatives at the NASA Ames Research Center.* Charles W. Wessner, ed. Washington, DC: National Academy Press.

National Research Council. 2001. *Trends in Federal Support of Research and Graduate Education.* Stephen A. Merrill, ed. Washington, DC: National Academy Press.

National Research Council. 2003. *Government-Industry Partnerships for the Development of New Technologies: Summary Report*. Charles W. Wessner, ed. Washington, DC: The National Academies Press.

National Research Council. 2003. *Securing the Future: Regional and National Programs to Support the Semiconductor Industry*. Charles W. Wessner, ed. Washington, DC: The National Academies Press.

National Research Council. 2004. *A Patent System for the 21st Century*. Washington, DC: The National Academies Press.

National Research Council. 2004. *Productivity and Cyclicality in Semiconductors: Trends, Implications, and Questions*. Dale W. Jorgenson and Charles W. Wessner, ed. Washington, DC: The National Academies Press.

National Research Council. 2004. *The Small Business Innovation Research Program: Program Diversity and Assessment Challenges*. Charles W. Wessner, ed. Washington, DC: The National Academies Press.

National Research Council. 2005. *Getting Up to Speed: The Future of Superconducting*. Susan L. Graham, Marc Snir, and Cynthia A. Patterson, eds. Washington, DC: The National Academies Press.

National Research Council. 2005. *Policy Implications of International Graduate Students and Post-doctoral Scholars in the United States*. Washington, DC: The National Academies Press.

National Research Council. 2007. *Enhancing Productivity Growth in the Information Age: Measuring and Sustaining the New Economy*. Dale W. Jorgenson and Charles W. Wessner, eds. Washington, DC: The National Academies Press.

National Research Council. 2007. *India's Changing Innovation System: Achievements, Challenges, and Opportunities for Cooperation*. Charles W. Wessner and Sujai J. Shivakumar, eds. Washington, DC: The National Academies Press.

National Research Council. 2007. *Innovation Policies for the 21st Century*. Charles W. Wessner, ed. Washington, DC: The National Academies Press.

National Research Council. 2007. *SBIR and the Phase III Challenge of Commercialization*. Charles W. Wessner, ed. Washington, DC: The National Academies Press.

National Research Council. 2008. *An Assessment of the SBIR Program*. Charles W. Wessner, ed. Washington, DC: The National Academies Press.

National Research Council. 2008. *An Assessment of the SBIR Program at the Department of Energy*. Charles W. Wessner, ed. Washington, DC: The National Academies Press.

National Research Council. 2008. *An Assessment of the SBIR Program at the National Science Foundation*. Charles W. Wessner, ed. Washington, DC: The National Academies Press.

National Research Council. 2008. *Innovative Flanders: Innovation Policies for the 21st Century*. Charles W. Wessner, ed. Washington, DC: The National Academies Press.

National Research Council. 2009. *21st Century Innovation Systems for Japan and the United States: Lessons from a Decade of Change*. Sadao Nagaoka, Masayuki Kondo, Kenneth Flamm, and Charles Wessner, eds. Washington, DC: The National Academies Press.

National Research Council. 2009. *An Assessment of the SBIR Program at the Department of Defense*. Charles W. Wessner, ed. Washington, DC: The National Academies Press.

National Research Council. 2009. *An Assessment of the SBIR Program at the National Institutes of Health*. Charles W. Wessner, ed. Washington, DC: The National Academies Press.

National Research Council. 2009. *An Assessment of the SBIR Program at the National Aeronautics and Space Administration*. Charles W. Wessner, ed. Washington, DC: The National Academies Press.

National Research Council. 2009. *Revisiting the Department of Defense SBIR Fast Track Initiative*. Charles W. Wessner, ed. Washington, DC: The National Academies Press.

National Research Council. 2009. *Understanding Research, Science, and Technology Parks: Global Best Practices*. Charles W. Wessner, ed. Washington, DC: The National Academies Press.

National Research Council 2010. *The Dragon and the Elephant, Understanding the Development of Innovation Capacity in China and India*. S. Merrill ed., Washington, DC: National Academies Press.

National Research Council. 2011. *The Future of Photovoltaics Manufacturing in the United States*. Charles W. Wessner, Rapporteur. Washington, DC: The National Academies Press.

National Research Council. Forthcoming. *Building the U.S. Battery Industry for Electric-Drive Vehicles: Progress, Challenges, and Opportunities*. Charles W. Wessner, Rapporteur. Washington, DC: The National Academies Press.

National Research Council. Forthcoming. *Clustering for 21st Century Prosperity*. Charles W. Wessner, Rapporteur. Washington, DC: The National Academies Press.

National Research Council. Forthcoming. *Growing Innovation Clusters for American Prosperity.* Charles W. Wessner, Rapporteur. Washington, DC: The National Academies Press.

National Science Board. 2004. *Science and Engineering Indicators 2004.* Arlington, VA: National Science Foundation.

National Science Board. 2006. *Science and Engineering Indicators 2006.* Arlington, VA: National Science Foundation.

National Science Board. 2008. *Science and Engineering Indicators 2008.* Arlington, VA: National Science Foundation.

National Science Foundation. 2004. *Science and Engineering Doctorate Awards: 2003.* NSF 05-300. Arlington, VA: National Science Foundation.

Needham, Joseph. 1954-1986. *Science and Civilization in China.* (Five volumes.) Cambridge: Cambridge University Press.

Nelson, R. R. 1993. *National Innovation Systems: A Comparative Analysis.* New York: Oxford University Press.

Nelson, R. R., and K. Nelson. 2002. "Technology, Institutions, and Innovation Systems." *Research Policy* 31:265-272.

Nelson, R. R., and N. Rosenberg. 1993. "Technical Innovation and National Systems." In *National Innovation Systems: A Comparative Analysis.* Richard R. Nelson, ed. Oxford, UK: Oxford University Press.

Nelson, R. R., and S. G. Winter. 1982. *An Evolutionary Theory of Economic Change.* Cambridge, MA: Harvard University Press.

Neuffer, John. 2010 "China: Intellectual Property Infringement, Indigenous Innovation Policies, and Frameworks for Measuring the Effects on the U.S. Economy." Testimony to the United States International Trade Commission Investigation June 15.

Nishimura, Kiyohiko, G., Takanobu Nakajima, and Kozo Kiyota. 2003. "Does the Natural Selection Mechanism Still Work in Severe Recessions? Examination of the Japanese Economy in the 1990s." *Journal of Economic Behavior and Organization* 58:53-78.

Odagari, H. 1999. "University-Industry Collaboration in Japan: Facts and Interpretations." In Lewis M. Branscomb, Fumio Kodama, and Richard Florida, eds. *Industrializing Knowledge: University-Industry Linkage in Japan and the United States.* Cambridge, MA: The MIT Press.

Odagiri, H., and A. Goto. 1993. "The Japanese System of Innovation: Past, Present and Future." In R. R. Nelson, ed. *National Systems of Innovation.* Oxford, UK: Oxford University Press.

Odagiri, H., and A. Goto. 1996. *Technology and Industrial Development in Japan: Building Capabilities by Learning, Innovation and Public Policy.* Oxford, UK: Oxford University Press.

Organisation for Economic Co-operation and Development. 1999. *Boosting Innovation: The Cluster Approach.* Paris, France: Organisation for Economic Co-operation and Development.

Organisation for Economic Co-operation and Development. 1999. *Managing National Innovation Systems.* Paris, France: Organisation for Economic Co-operation and Development.

Organisation for Economic Co-operation and Development. 2001. *The New Economy: Beyond the Hype—The OECD Growth Project.* Paris, France: Organisation for Economic Co-operation and Development.

Organisation for Economic Co-operation and Development. 2001. *Social Sciences and Innovation.* Washington, DC: Organisation for Economic Co-operation and Development.

Organisation for Economic Co-operation and Development. 2004. *Summary Report: Micro-policies for Growth and Productivity.* DSTI/IND(2004)7. Paris, France: Organisation for Economic Co-operation and Development. October.

Organisation for Economic Co-Operation and Development. 2008. *OECD Reviews of Innovation Policy: China.* Paris, France: Organisation for Economic Co-operation and Development.

Organisation for Economic Co-operation and Development. 2011. *Main Science and Technology Indicators.* Paris, France: Organisation for Economic Co-operation and Development.

Orszag, Peter, and Thomas Kane. 2003. "Funding Restrictions at Public Universities: Effects and Policy Implications." Brookings Institution Working Paper. September.

Oughton, Christine. 1997. "Competitiveness in the 1990s." *The Economic Journal* 107(444): 1486-1503.

Oughton, Christine, Mikel Landabaso, and Kevin Morgan. 2002. "The Regional Innovation Paradox: Innovation Policy and Industrial Policy." *The Journal of Technology Transfer* 27(1).

Owens, William, with Edward Offley. 2001. Lifting the Fog of War. Baltimore, MD: Johns Hopkins University Press. Park, W. G., and J. C. Ginarte. 1997. "Intellectual Property Rights and Economic Growth." *Contemporary Economic Policy* XV(July):51-61.

Patel, P., and K. Pavitt. 1994. "National Innovation Systems: Why They are Important and How They Might Be Compared?" *Economic Change and Industrial Innovation.*

Peck, M. J., R. C. Levin, and Akira Goto. 1988. "Picking Losers: Public Policy Toward Declining Industries in Japan." In J. B. Shoven, ed. *Government Policy Toward Industry in the United States and Japan.* Cambridge, UK: Cambridge University Press. Pp. 165-239.

Peek, Joe, and Eric S. Rosengren. 2005. "Unnatural Selection: Perverse Incentives and the Misallocation of Credit in Japan." *The American Economic Review* 95(4):114-1166.

Penrose, E. 1951. *The Economics of the International Patent System.* Baltimore, MD: Johns Hopkins University Press.

Perez, Carlota. 2002. *Technological Revolutions and Financial Capital.* Cheltenham, UK: Edward Elgar.

Phan, Phillip H., and Donald S. Siegel. 2006. "The Effectiveness of University Technology Transfer: Lessons Learned from Qualitative and Quantitative Research in the U.S. and U.K." Rensselaer Working Papers in Economics 0609. Troy, NY: Rensselaer Polytechnic Institute Department of Economics.

Phan, Phillip H., Donald S. Siegel, and Mike Wright. 2005. "Science Parks and Incubators: Observations, Synthesis and Future Research." *Journal of Business Venturing* 20(2):165-182.

Phillimore, J. 1999. "Beyond the Linear View of Innovation in Science Park Evaluation: An Analysis of Western Australian Technology Park." *Technovation* 19(11):673 680.

Polenske, Karen, Nicolas Rockler, et al. 2004. *Closing the Competitive Gap: A Retrospective Analysis of the ATP 2mm Project.* NIST GCR 03-856. Gaithersburg, MD: National Institute of Standards and Technology.

Porter, M. E., 1998. "Clusters and the New Economics of Competition." *Harvard Business Review* 76(6):77-90.

Porter, M. E., and M. Sakakibara. 2004. "Competition in Japan." *The Journal of Economic Perspectives* 18(1).

Posen, Adam. 1998. *Restoring Japan's Economic Growth.* Washington, DC: Peterson Institute for International Economics.

Posen, Adam S. 2001. "Japan." In Benn Steil, David G. Victor, and Richard R. Nelson, eds. *Technological Innovation and Economic Performance.* Princeton, NJ: Princeton University Press.

Powell, Jeanne, and Francisco Moris. 2002. *Different Timelines for Different Technologies.* NISTIR 6917. Gaithersburg, MD: National Institute of Standards and Technology.

President's Council of Advisors on Science and Technology. 2004. "Sustaining the Nation's Innovation Ecosystems." Washington, DC: Executive Office of the President. January.

President's Council of Advisors on Science and Technology. 2004. "Sustaining the Nation's Innovation System: Report on Information Technology Manufacturing and Competitiveness." Washington, DC: Executive Office of the President. January.

President's Information Technology Advisory Committee. 2005. "Cybersecurity: A Crisis of Prioritization." Report to the President. February.

PricewaterhouseCoopers. 2006. "China's Impact on the Semiconductor Industry: 2005 Update." New York: PricewaterhouseCoopers.

Procassini, Andrew. 1995. Competitors in Alliance: Industry Associations, Global Rivalries, and Business-Government Relations. New York: Greenwood Publishing.

Purvis, G. 2002. "Moving into the Real World." *Electronic Business.* July 1.

Qiang, Christine Zhen-Wei. 2009. "Broadband Infrastructure Investment in Stimulus Packages: Relevance for Developing Countries," Washington DC: World Bank.

Raduchel, William. 2006. "The End of Stovepiping." In National Research Council. *The Telecommunications Challenge: Changing Technologies and Evolving Policies.* Charles W. Wessner, ed. Washington, DC: The National Academies Press.

Reid, T. R. 2004. *The United States of Europe: The New Superpower and the End of American Supremacy.* New York: Penguin Press.

Reiko, Yamada. 2001. "University Reform in the Post-massification Era in Japan: Analysis of Government Education Policy for the 21st Century." *Higher Education Policy* 14(4).

Reuters. 2006. "China Sees No Quick End to Economic Boom." February 21.

Rich, Ben. 1996. *Skunkworks.* New York: Back Bay Books.

Rodrik, Dani and Arvind Subramanian. 2004. "From 'Hindu Growth' to Productivity Surge: The Mystery of the Indian Growth Transition." NBER Working Paper 10376.

Romanainen, Jari. 2001. "The Cluster Approach in Finnish Technology Policy." Pp. 377-388 in Edward M. Bergman, Pim den Hertog, and David Charles, eds. *Innovative Clusters: Drivers of National Innovation Systems.* OECD Proceedings. Washington, DC: Organisation for Economic Co-operation and Development.

Romer, Paul. 1990. "Endogenous Technological Change." *Journal of Political Economy* 98:72-102.

Rosenberg, N., and R. Nelson. 1994. "American Universities and Technical Advance in Industry." *Research Policy* 23:325-348.

Rosenzweig, R., and B. Turlington. 1982. *The Research Universities and Their Patrons*. Berkeley, CA: University of California Press.

Rothaermel, F. T., and M. C. Thursby. 2005. "Incubator Firm Failure or Graduation? The Role of University Linkages." *Research Policy* 34(7):1076-1090.

Rothaermel, F. T., and M. C. Thursby. 2005. "University-Incubator Firm Knowledge Flows: Assessing Their Impact on Incubator Firm Performance." *Research Policy* 34(3):302-320.

Rowen, Henry S., and A. Maria Toyoda. 2002. "From Kiretsu to Start-ups: Japan's Push for High Tech Entrepreneurship." Asia-Pacific Research Center Working Paper. Stanford, CA.

Ruegg, Rosalie, and Irwin Feller. 2003. *A Toolkit for Evaluating Public R&D Investment: Models, Methods, and Findings from ATP's First Decade*. NIST GCR 03-857. Gaithersburg, MD: National Institute of Standards and Technology.

Ruttan, Vernon W. 2002. *Technology, Growth and Development: An Induced Innovation Perspective*. Oxford, UK: Oxford University Press.

Ruttan, Vernon W. 2006. *Is War Necessary for Economic Growth, Military Procurement and Technology Development?* New York: Oxford University Press.

Ruttan, Vernon W. 2006. "Will Government Programs Spur the Next Breakthrough?" *Issues in Science and Technology* Winter.

Rutten, Roel, and Frans Boekema. 2005. "Innovation, Policy and Economic Growth: Theory and Cases." *European Planning Studies* 13(8).

Rycroft, Robert W., and Don E. Kash. 1999. "Innovation Policy for Complex Technologies." *Issues in Science and Technology* Fall.

Saito, Ken. 1987. *Research on a New Concern: RIKEN Industrial Group*. (In Japanese.) Jichosha

Sakakibara, M., and L. Branstetter. 2001. "Do Stronger Patents Induce More Innovation? Evidence from the 1988 Japanese Patent law Reforms." *Rand Journal of Economics* 32:77-100.

Sakakibara, M., and L. Branstetter. 2002. *Measuring the Impact of ATP-Funded Research Consortia on Research Productivity of Participating Firms, A Framework Using Both U.S. and Japanese Data*. NIST GCR 02-830. Gaithersburg, MD: National Institute of Standards and Technology.

Sakamoto, Kozo, and Masayuki Kondo. 2004. "The Analysis of University-Industry Research Collaborations by Time Series and Corporate Characteristics." (In Japanese.) *Development Engineering* 10:11-26.

Sallet, J, J. Masterman, and E. Paisley. 2009. "The Geography of Innovation." Washington, DC: Center for American Progress.

Sapolsky, Harvey M. 1990. *Science and the Navy—The History of the Office of Naval Research*. Princeton, NJ: Princeton University Press.

Saxenian, AnnaLee. 1994. *Regional Advantage: Culture and Competition in Silicon Valley and Route 128*. Cambridge, MA: Harvard University Press.

Schaffer, Teresita. 2002. "Building a New Partnership with India." *Washington Quarterly* 25(2):31-44..

Scherer, F. M. 2001. "U.S. Government Programs to Advance Technology." *Revue d'Economie Industrielle* 0(94):69-88.

Science Committee of the Council for Science and Technology of the Ministry of Education, Culture, Sports, Science and Technology. 2005. *Sciences Policy to Support Diversity in Research*. (Kenkyuu-no-tayousei wo-sasaeru Gakujutu-seisaku). October 13.

Scotchmer, Suzanne. 1996. "Protecting Early Innovators: Should Second-Generation Products Be Patentable?" *Rand Journal of Economics* 27:322-331.

Scotchmer, Suzanne. 2004. *Innovation and Incentives*. Cambridge, MA: The MIT Press.

Segal, Adam. 2010. China's Innovation Wall; Beijing's Push for Homegrown Technology, *Foreign Affairs*.

Shane, S. 2004. *Academic Entrepreneurship*. Cheltenham, UK: Edward Elgar Publishing.

Shang, Yong. 2006. "Innovation: New National Strategy of China." Presentation at Industrial Innovation in China. Levin Institute Conference. July 24-26.

Shanghai Municipal Government. 2004. Notice of the Shanghai Municipal Government Regarding Distributing the Outline of Shanghai's Intellectual Property Strategy (2004-2010). September 14.

Shapiro, Carl. 2001. "Navigating the Patent Thicket: Cross License, Patent Pools and Standard-Setting." In Adam Jaffe, Joshua Lerner, and Scott Stern, eds. *Innovation Policy and the Economy*. Cambridge, MA: The MIT Press.

Shearmur, R., and D. Doloreux. 2000. "Science Parks: Actors or Reactors? Canadian Science Parks in their Urban Context." *Environment and Planning* 32(6):1065-1082.

Sheehan, Jerry, and Andrew Wyckoff. 2003. "Targeting R&D: Economic and Policy Implications of Increasing R&D Spending." DSTI/DOC(2003)8. Paris, France: Organisation for Economic Co-operation and Development.

Shenzhen Daily. 2005. "Nation May Introduce Antimonopoly Law." December 30.

Sherwin, Martin, and Kai Bird. 2005. *American Prometheus: The Triumph and Tragedy of J. Robert Oppenheimer*. New York: Alfred A. Knopf.

Shin, Roy W. 1997. "Interactions of Science and Technology Policies in Creating a Competitive Industry: Korea's Electronics Industry." *Global Economic Review* 26(4):3-19.

Siegel, Donald S., David Waldman, and Albert Link. 2004. "Toward a Model of the Effective Transfer of Scientific Knowledge from Academicians to Practitioners: Qualitative Evidence from the Commercialization of University Technologies." *Journal of Engineering and Technology Management* 21(1-2):115-142.

Siegel, Donald S., P. Westhead, and M. Wright. 2003. "Assessing the Impact of Science Parks on Research Productivity: Exploratory Firm-Level Evidence from the United Kingdom." *International Journal of Industrial Organization* 21(9):1357-1369.

Simon, Dennis and Cong Cao. 2009. *China's Emerging Technological Edge: Addressing the Role of High-End Talent*. Cambridge: Cambridge University Press.

Skolnikoff, Eugene B. 1993. "Knowledge Without Borders? Internationalization of the Research Universities." *Daedalus* 122(4).

Smith, Kathlin. 1998. *The Role of Scientists in Normalizing U.S.-China Relations: 1965-1979*. Council on Library and Information Resources

Smits, Ruud, and Stefan Kuhlmann. 2004. "The Rise of Systemic Instruments in Innovation Policy." *International Journal of Foresight and Innovation Policy* 1(1/2).

Soete, Luc G., and Bastiann J. ter Weel. 1999. "Innovation, Knowledge Creation and Technology Policy: The Case of the Netherlands." *De Economist* 147(3). September.

Sofouli, E., and N. S. Vonortas. 2007. "S&T Parks and Business Incubators in Middle-Sized Countries: The Case of Greece." *Journal of Technology Transfer* 32(5):525-544.

Solow, Robert. 2000. *Growth Theory: An Exposition*. New York: Oxford University Press. 2nd edition.

Solow, Robert. 2000."Toward a Macroeconomics of the Medium Run." *Journal of Economic Perspectives* Winter.

Spence, Michael. 1974. *Market Signaling: Informational Transfer in Hiring and Related Processes.* Cambridge, MA: Harvard University Press.

Spencer, William, and T. E. Seidel. 2004. "International Technology Roadmaps: The U.S. Semiconductor Experience." In National Research Council. *Productivity and Cyclicality in Semiconductors: Trends, Implications, and Questions.* Dale W. Jorgenson and Charles W. Wessner, eds. Washington, DC: The National Academies Press.

Spencer, W. J., L. Wilson, and R. Doering. 2004. "The Semiconductor Technology Roadmap." *Future Fab International* 18.

Springut, Micah, et al. 2011. "China's Program for Science and Technology Modernization: Implications for American Competitiveness." U.S.-China Economic and Security Review Commission. January 2011.

Stanford University. 1999. *Inventions, Patents and Licensing: Research Policy Handbook.* Document 5.1. July 15.

State Council of the Peoples' Republic of China. 2006. *Outline of the National Medium- and Long-term Program on Scientific and Technological Development (2006-2020).* February 9.

Statistics Bureau, Ministry of Internal Affairs and Communications. 2005. *Report on the Survey of Research and Development 2004.* Ministry of Internal Affairs and Communications.

Stephan, P. 2001. "Educational Implication of University-Industry Technology Transfer." *Journal of Technology Transfer* 26:199-205.

Sternberg, R. 1990. "The Impact of Innovation Centres on Small Technology-based Firms: The Example of the Federal Republic of Germany." *Small Business Economics* 2(2):105-118.

Stiglitz, Joseph. 2005. "The Ethical Economist." *Foreign Affairs.* Council on Foreign Affairs. November/December. Accessed at <http:www.foreignaffairs.org>.

Stokes, Donald E. 1997. *Pasteur's Quadrant: Basic Science and Technological Innovation.* Washington, DC: Brookings Institution Press.

Strout, E. 2005. "Gift of a Book Was a Key to Intel Founder's Big Donation to City College of New York." *Chronicle of Higher Education.* December 2. P. A27.

Su, Yun-Shan, and Ling-Chun Hung. 2008. "Spontaneous vs. Policy-driven: The Origin and Evolution of the Biotechnology Cluster." *Technological Forecast and Social Change.*

Swann, G. M. P. 1998. "Towards a Model of Clustering in High-Technology Industries." In G. M. P. Swann, M. Prevezer, and D. Stout, eds. *The Dynamics of Industrial Clustering*. Oxford, UK: Oxford University Press.

Talele, Chitram J. 2003. "Science and Technology Policy in Germany, India and Pakistan." *Indian Journal of Economics and Business* 2(1):87-100.

Tanaka, Nobua. 2005. Presentation at the International Forum on Technology Foresight and National Innovation Strategies. Seoul, Republic of Korea. November 4.

Tassey, G. 1997. *The Economics of R&D Policy*. Westport, CT: Quorum Books.

Tassey, G. 2004. "Policy Issues for R&D Investment in a Knowledge-based Economy." *Journal of Technology Transfer* 29:153-185.

Tassey, G. 2007. *The Innovation Imperative*. Cheltenham, UK: Edward Elgar.

Teubal, Morris. 2002. "What Is the Systems Perspective to Innovation and Technology Policy and How Can We Apply It to Developing and Newly Industrialized Economies?" *Journal of Evolutionary Economics* 12(1-2).

Thelin, J. 2004. *A History of American Higher Education*. Baltimore, MD: Johns Hopkins University Press.

Thomas, J. R., and W. H. Schacht. 2008. "Patent Reform in the 110th Congress: Innovation Issues." Report for Congress. Washington, DC: Congressional Research Service. Order Code RL33996.

Thursby, J., and M. Thursby. 2002. "Who is Selling the Ivory Tower? Sources of Growth in University Licensing." *Management Science* 48:90-104.

Thursby, J., and M. Thursby. 2004. "Industry Perspectives on Licensing University Technologies: Sources and Problems." *AUTM Journal* P. 2000.

U.S.-China Joint Commission on Commerce and Trade (JCCT). 2005. *Outcomes on Major U.S. Trade Concerns*. Washington, DC: The Office of the United States Trade Representative.

U.S. Congress. 1988. *Omnibus Trade and Competitiveness Act of 1988* (P.L. 100-418, codified in 15 U.S.C. 278n.) and later amended by the *American Technology Preeminence Act of 1991* (P.L. 102-245, codified in 15 U.S.C. 3701).

U.S. Department of Commerce. 2008. *Innovation Measurement: Tracking the State of Innovation in the American Economy, Report to the Secretary.* Washington, DC: U.S. Department of Commerce. January. Available at <*http://www.innovationmetrics.gov/Innovation%20Measurement%20 01-08.pdf*>.

U.S. General Accounting Office. 2002. *Export Controls: Rapid Advances in China's Semiconductor Industry Underscore Need for Fundamental U.S. Policy Review.* GAO-020620. Washington, DC: U.S. General Accounting Office. April.

U.S. Small Business Administration. 2004. "Small Business by the Numbers." Office of Advocacy. Washington, DC: U.S. Small Business Administration. June.

Vaidyanathan, G. 2008. "Technology Parks in a Developing Country: The Case of India." *Journal of Technology Transfer* 33(3):285-299.

Van Atta, Richard. 2004. "Energy and Climate Change Research and the DARPA Model." Presentation to the Washington Roundtable on Science and Public Policy. November 3.

Van Atta, Richard. 2008. "Fifty Years of Innovation and Discovery." In *DARPA: 50 Years of Bridging the Gap.* Arlington, VA: Defense Advanced Research Projects Agency. April.

Van Atta, Richard, et al. 1991. *DARPA Technical Accomplishments.* Volumes I-V. Alexandria, VA: Institute for Defense Analysis.

Van Atta, Richard, et al. 1991. *DARPA Technological Accomplishments: An Historical Review of Selected DARPA Projects.* Alexandria, VA: Institute for Defense Analysis.

Van Atta, Richard, and Michael Lippitz. 2003. *Transformation and Transition: DARPA's Role in Fostering an Emerging Revolution in Military Affairs.* Volume 1: Overall Assessment. Alexandria, VA: Institute for Defense Analysis.

Van Looy, Bart, K. Debackere, and T. Magerman. 2005. *Assessing Academic Patent Activity: The Case of Flanders.* Leuven, Belgium: SOOS.

Van Looy, Bart, Marina Ranga, Julie Callaert, Koenraad Debackere, and Edwin Zimmermann. 2004. "Combining Entrepreneurial and Scientific Performance in Academia: Towards a Compounded and Reciprocal Matthew-effect?" *Research Policy* 33(3):425-441.

Vedovello, C. 1997. "Science Parks and University-Industry Interaction: Geographical Proximity between the Agents as a Driving Force." *Technovation* 17(9):491-502.

Vervliet, Greta. 2006. *Science, Technology, and Innovation*. Brussels, Belgium: Ministry of Flanders, Science and Innovation Administration.

Vest, C. 2005. "Industry, Philanthropy and Universities-The Roles and Influences of the Private Sector in Higher Education." 2005 Clark Kerr Lecture. University of California at Berkeley. September 13.

Veugelers, Reinhilde, Jan Larosse, Michele Cincera, Donald Carchon, and Roger Kalenga-Mpala. 2004. "R&D Activities of the Business Sector in Flanders: Results of the R&D Surveys in the Context of the 3% Target." *IWT-Studies* (46). Brussels, Belgium.

Wadhwa, Vivek, and Gary Gereffi. 2005. *Framing the Engineering Outsourcing Debate*. Durham, NC: Duke University.

Walcott, Susan M. 2003. *Chinese Science and Technology Industrial Parks*. Aldershot, UK: Ashgate Publishing.

Waldrop, M. Mitchell. 1992. *Complexity: The Emerging Science at the Edge of Order and Chaos*. New York: Simon & Schuster.

Waldrop, M. Mitchell. 2001. *The Dream Machine*. New York: Viking Press.

Walsh, J., A. Arora, and W. Cohen. 2003. "Research Tool Patent and Licensing and Biomedical Innovation." In *Patents in the Knowledge-Based Economy*. W. Cohen and S. Merrill, eds. Washington, DC: The National Academics Press.

Walsh, J. P., and W. M. Cohen. 2004. "Does the Golden Goose Travel? A Comparative Analysis of the Influence of Public Research on Industrial R&D in the U.S. and Japan." Mimeo.

Wang, Changyong. 2005. "IPR Sails Against Current Stream." *Caijing Magazine* October 17. Available online at <http://caijing.hexun.com>.

Wang, Liwei. 1993. "The Chinese Traditions Inimical to the Patent Law." *Northwestern Journal of International Lawand Commerce* Fall.

Warsh, David. 2006. *Knowledge and the Wealth of Nations*. New York: W. W. Norton. The Washington Post. 2006. "Chinese to Develop Sciences, Technology." February 10. P. A16.

Wen, Jiang, and Shinichi Kobayashi. 2001. "Exploring Collaborative R&D Network: Some New Evidence in Japan." *Research Policy* 30:1309-1319.

Wessner, Charles W. 2005. "Entrepreneurship and the Innovation Ecosystem." In David B. Audretsch, Heike Grimm, and Charles W. Wessner, eds. *Local Heroes in the Global Village: Globalization and the New Entrepreneurship Policies*. New York: Springer.

Westhead, P. 1995. "New Owner-Managed Businesses in Rural and Urban Areas in Great Britian: A Matched Pairs Comparison." *Regional Studies* 29(4):367-380.

Westhead, P. 1997. "R&D 'Inputs' and 'Outputs' of Technology-based firms Located On and Off Science Parks." *R&D Management* 27(1):45-61. *Kingdom*. London, UK: HMSO.

Westhead, P., and D. Storey. 1994. *An Assessment of Firms Located On and Off Science Parks in the United*

Westhead, P., and D. Storey. 1997. "Financial Constraints on the Growth of High-Technology Small Firms in the U.K." *Applied Financial Economics* 7(2):197-201.

Westhead, P., D. J. Storey, and M. Cowling. 1995. "An Exploratory Analysis of the Factors Associated with the Survival of Independent High-Technology Firms in Great Britain." In F. Chittenden, M. Robertson, and I. Marshall, eds. *Small Firms: Partnerships for Growth*. London, UK: Paul Chapman.

Westhead, P., and M. Cowling. 1995. "Employment Change in Independent Owner-Managed High-Technology Firms in Great Britain." *Small Business Economics* 7(2):111-140.

Westhead, P., and S. Batstone. 1998. "Independent Technology-based Firms: The Perceived Benefits of a Science Park Location." *Urban Studies* 35(12):2197-2219. The White House 2009. "Remarks by President Obama," Prague: April 5

Wiener, Norbert. 1948. *Cybernetics or Control and Communication in the Animal and the Machine*. Cambridge, MA: The MIT Press.

World Bank. 2004. *Innovation Systems: World Bank Support of Science and Technology Development*. Vinod Kumar Goel, ed. Washington, DC: World Bank.

World Bank. 2006. *The Environment for Innovation in India. South Asia Private Sector Development and Finance Unit*. Washington, DC: World Bank.

World Bank International Finance Corporation. 2006. *Doing Business in 2006: Creating Jobs*. Washington, DC: International Bank for Reconstruction and Development.

World Trade Organization Working Party on the Accession of China. 2001. *Report of the Working Party on the Accession of China*. WT/MN(01)/3. November 10.

Wu, Ching. 2006. "China to Build 30 New Science and Technology Parks." *SciDev.net*. April 19.

Yan, Dai. 2004. "Anti-Monopoly Legislation on the Way." *China Daily* June 18.

Yang, Lei, ed. 2006. "Chinese WAPI Delegation Quits Prague Meeting." *Xinhua*. June 8. Available at *<http://news.xinhuanet.com>*.

Yasuda, T. 2005. "Seisakukinyuu no riyou" (The Utilization of Public Finance). In Nippon no shinki kaigyou kigyou (Startup Enterprises in Japan). K. Kustuna and T. Yasuda, ed. Hakutousha.

Yoshida, Fujio. 1967. "Preparation of Legal System for Capital Liberalization (Part 3)." Panel Discussion in Zaikei Shoho. July 17.

Yoshida, Junko. 2006. "Grenoble Lure: Un-French R&D." *EE Times* June 12.

Young, John. 2007. Info Memo for Secretary of Defense Robert M. Gates. DoD Science and Technology Program. August 24.

Yukio, Sato. 2001. "The Structure and Perspective of Science and Technology Policy in Japan." In Phillipe Laredo and Phillipe Mustar, eds. *Research and Innovation Policies in the New Global Economy: An International Comparative Analysis*. Cheltenham, UK: Edward Elgar.

Zachary, G. Pascal. 1999. *Endless Frontier: Vannevar Bush, Engineer of the American Century*. Cambridge, MA: The MIT Press.

Zeigler, Nicholas J. 1997. *Governing Ideas: Strategies for Innovation in France and Germany*. Ithaca, NY: Cornell University Press.

Zemin, Jiang, General Secretary of the Communist Party of China Central Committee. 1999. Keynote Speech at the National Technological Innovation Conference, August 23.

Ziedonis, R. H., and B. H. Hall. 2001. "The Effects of Strengthening Patent Rights on Firms Engaged in Cumulative Innovation: Insights from the Semiconductor Industry." In Gary Libecap, ed., *Entrepreneurial Inputs and Outcomes: New Studies of Entrepreneurship in the United States*. Volume 13 of *Advances in the Study of Entrepreneurship, Innovation, and Economic Growth*. Amsterdam, The Netherlands: Elsevier Science.

Zhang, Chunlin, Douglas Zhihua Zeng, William Peter Mako, and James Seward. 2009. *Promoting Enterprise-Led Innovation in China, 2009*. Washington, D. C.: The International Bank for Reconstruction and Development/The World Bank,

Zheng, Liang and Xue, Lan. 2010 "The evolution of China's IPR system and its impact on the patenting behaviours and strategies of multinationals in China." Volume 51 of *International Journal of Technology Management*.